I0066180

Aircraft Manufacturing, Safety and Control

*Edited by Melih Cemal Kuşhan
and Seyid Fehmi Diltemiz*

Published in London, United Kingdom

Aircraft Manufacturing, Safety and Control
http://dx.doi.org/10.5772/intechopen.1006780
Edited by Melih Cemal Kuşhan and Seyid Fehmi Diltemiz

Contributors
Frances Brazier, Jean-Charles Chaudemar, Melih Cemal Kushan, Pieter van Langen, Rob Vingerhoeds, Seref Demirci, Seyid Fehmi Diltemiz, Sophie Lemoussu, Wafa Ben Hassen, Ümit Deniz Göker

© The Editor(s) and the Author(s) 2025

The rights of the editor(s) and the author(s) have been asserted in accordance with the Copyright, Designs and Patents Act 1988. All rights to the book as a whole are reserved by INTECHOPEN LIMITED. The book as a whole (compilation) cannot be reproduced, distributed or used for commercial or non-commercial purposes without INTECHOPEN LIMITED's written permission. Enquiries concerning the use of the book should be directed to INTECHOPEN LIMITED rights and permissions department (permissions@intechopen.com).

Violations are liable to prosecution under the governing Copyright Law.

(cc) BY

Individual chapters of this publication are distributed under the terms of the Creative Commons Attribution 4.0 License which permits commercial use, distribution and reproduction of the individual chapters, provided the original author(s) and source publication are appropriately acknowledged. If so indicated, certain images may not be included under the Creative Commons license. In such cases users will need to obtain permission from the license holder to reproduce the material. More details and guidelines concerning content reuse and adaptation can be found at http://www.intechopen.com/copyright-policy.html.

Notice

Statements and opinions expressed in the chapters are these of the individual contributors and not necessarily those of the editors or publisher. No responsibility is accepted for the accuracy of information contained in the published chapters. The publisher assumes no responsibility for any damage or injury to persons or property arising out of the use of any materials, instructions, methods or ideas contained in the book.

First published in London, United Kingdom, 2025 by IntechOpen
IntechOpen is the global imprint of INTECHOPEN LIMITED, registered in England and Wales, registration number: 11086078, 167-169 Great Portland Street, London, W1W 5PF, United Kingdom

For EU product safety concerns: IN TECH d.o.o., Prolaz Marije Krucifikse Kozulić 3, 51000 Rijeka, Croatia, info@intechopen.com or visit our website at intechopen.com.

British Library Cataloguing-in-Publication Data
A catalogue record for this book is available from the British Library

Aircraft Manufacturing, Safety and Control
Edited by Melih Cemal Kuşhan and Seyid Fehmi Diltemiz
p. cm.
Print ISBN 978-1-83634-941-9
Online ISBN 978-1-83634-940-2
eBook (PDF) ISBN 978-1-83634-942-6

If disposing of this product, please recycle the paper responsibly.

IntechOpen

intechopen.com

Built by scientists, for scientists

Explore all IntechOpen books

Meet the editors

Melih Cemal Kushan, Ph.D., is a Professor in the Department of Aeronautical Engineering at Eskisehir Osmangazi University. He has a B. Sc. degree in Mechanical Engineering from Dokuz Eylul University, Izmir, Turkey; an M. Sc. degree in Industrial Engineering from Anadolu University, Eskisehir, Turkey; and a Ph.D. degree in Mechanical Engineering from Eskisehir Osmangazi University, Eskisehir, Turkey. He conducted his postdoctoral research in the College of Science and Engineering, Department of Mechanical Engineering at James Cook University, Australia. Professor Kushan has more than 270 publications on aviation and defense technology, and his publications have been cited in more than 3350 scientific studies. He has completed 22 research projects in the field of materials science and manufacturing of aerospace components. He is a member of the Society of Automobile Engineers (SAE International), the Eskisehir National Aviation Cluster (ESAC), and the Turkish Chamber of Architects and Engineers (TMMOB). For his academic studies, he has been in more than 20 countries, including Wollongong University in Australia, International Turkmen Turkish University in Turkmenistan, and Vilnius Gediminas Technical University in Lithuania. The awards he has received are; Best Paper Award/Polyteks Award, 3rd International Symposium on Fiber and Polymer Research, Bursa, Turkey; Best Paper Award, IAARHIES 28th International Conference on Engineering & Technology ICET – 2016, Shanghai, China; Excellent Presentation Award, 4th Asia Conference on Mechanical and Materials Engineering (ACMME 2016), Kuala Lumpur, Malaysia; VI. Necdet Eraslan Project Competition 3rd Prize, 2016, Istanbul, Turkey; TTGV Project Incentive Award, 2009, Ankara, Turkey; and many Publication Performance Awards from Eskisehir Osmangazi University, Eskisehir, Turkey.

Seyid Fehmi Diltemiz was born in Malatya in 1973. He completed his undergraduate education in the Metallurgical and Materials Engineering Department at Yıldız Technical University in 1994. He received his Ph.D. degree from the Metallurgical Engineering Department at Eskisehir Osmangazi University in 2010. Between 1999 and 2020, he worked at the Metallurgy Laboratory of the Eskisehir 1st Air Supply and Maintenance Center Command. He currently serves as an Assistant Professor in the Aeronautical Engineering Department at Eskisehir Osmangazi University.

Contents

Preface

Aviation advances when engineering disciplines meet along a single production line and a single flight path. This edited volume, *Aircraft Manufacturing, Safety and Control*, was conceived to bring those disciplines together in one place for readers who design, build, certify, and operate aeroplanes. Our aim here is not to present new findings, but to provide a coherent gateway to the themes that the contributed chapters explore in depth across manufacturing science, airworthiness, and flight/control systems.

The book opens with context: the industrial and regulatory frameworks within which aircraft are conceived, the roles of design assurance, configuration management, and quality systems, and how certification requirements shape every downstream decision in manufacturing and testing. Readers new to the field will find an accessible language for concepts that often sit behind specialist acronyms and internal procedures.

Manufacturing content then traces the path from materials to structure to assembly. The chapters examine metallic and composite process routes, including machining, forming, lay-up and curing, bonding and fastening, tolerance management, and producibility. Non-destructive inspection, repairability, and factory-floor digitalization are introduced not as afterthoughts, but as design drivers that influence weight, cost, and schedule. Sustainability is a continuous thread: material and energy efficiency, process waste reduction, and maintainability are discussed as practical levers rather than abstract goals.

Safety is treated as an engineering system, not merely a compliance outcome. The contributions set out the logic of safety assessment, reliability thinking, and damage tolerance, show how hazards are identified and mitigated across the lifecycle, and explain how maintenance practices and continued airworthiness feed back into design. Human factors, organizational learning, and data-driven monitoring complement the classical methods. Throughout, the message is consistent: safety emerges from decisions made early, disciplined processes during build, and honest measurement in service.

Control connects design intent to operational reality. The chapters addressing guidance, navigation, and control explain how modern architectures blend classical and state-space thinking with robust implementation and verification. Topics such as fly-by-wire, fault detection and isolation, health monitoring, and autonomy are linked explicitly to manufacturing constraints and safety assurance. In this way, the volume shows that "manufacturing, safety and control" are not separate domains but interdependent elements of a single engineering practice.

This volume is intended for a broad readership: engineers in training, specialists who wish to refresh fundamentals outside their own niche, and managers who must integrate technical, regulatory, and economic viewpoints. Each chapter is written to be self-contained and practical, with definitions, process insights, and worked examples that readers can translate to their day-to-day work.

We are grateful to the authors for their professionalism and patience throughout the editorial process, and to the reviewers for their rigorous and constructive feedback, which strengthened every contribution. We also thank the production team for their careful handling of the manuscript, and our students and colleagues for the many questions that helped us shape the scope and emphasis of this book. In particular, we would like to acknowledge Marijana Josipovic, Commissioning Editor, and Josip Knapic, Publishing Process Manager, for their steady guidance and support throughout the process.

We hope this volume serves as a valuable reference and an invitation: a reference for doing the present work well, and an invitation to continue improving how we design, manufacture, assure, and control the aeroplanes that connect people and ideas.

Melih Cemal Kushan
Aerospace Technologies Application and Research Center,
Eskişehir Osmangazi University,
Eskişehir, Türkiye

Seyid Fehmi Diltemiz
Aeronautical Engineering Department,
Eskişehir Osmangazi University,
Eskişehir, Türkiye

Chapter 1

Designing Aircraft in a Way that Reduces Human Error

Seref Demirci

Abstract

The causes of most incidents and accidents are attributed to humans, even though they are based on aircraft design and configuration. This chapter emphasizes the critical need for error-proof design in aviation to reduce human errors in aircraft maintenance and operation. Traditional models often assume human reliability as flawless, overlooking the potential for human error in complex systems. Findings reveal that many incidents and accidents stem from design-related issues rather than human shortcomings, suggesting that designing aircraft with error-proofing principles could significantly improve safety. This approach advocates for defining error-proofing levels, from basic awareness to advanced prevention, based on the severity of potential errors. Methods like Poka-Yoke and Murphy's law applications enhance system resilience by minimizing opportunities for human error. The adoption of error-proofing standards as part of the aviation safety framework could lead to fewer incidents and operational disruptions, such as in-flight shutdowns and delays. The study calls for global aviation authorities to recognize the role of human reliability in system design and to incorporate comprehensive error-proofing standards. By shifting focus from blaming individuals to proactive design solutions, this approach aims to promote safer and more efficient flight operations across the industry.

Keywords: human reliability, human-machine interface, human errors, Poka-Yoke, error-proof design, total system approach

1. Introduction

As a result of most incident and accident investigations, the majority of the errors are thought to be caused by human error. The investigation conclusions must be based on absolute facts to eradicate the mistakes effectively. According to the numbers given in an aviation magazine published by Boeing [1], around 80% of airplane accidents are caused by human error and 20% are machinery malfunctions. Unfortunately, in investigations, the causes of most accidents are attributed to humans, even though they are based on aircraft design and configuration. For example, the Boeing 737 NG crash near Amsterdam in 2009 and the two 737 MAX crashes recently in Indonesia and Ethiopia were caused by malfunctioning computer systems that rely on data from a single sensor, as shown in **Figures 1** and **2** [2]. In systems where computer automation replaces the pilot's role, full automation [2], the performance of the automation

IntechOpen

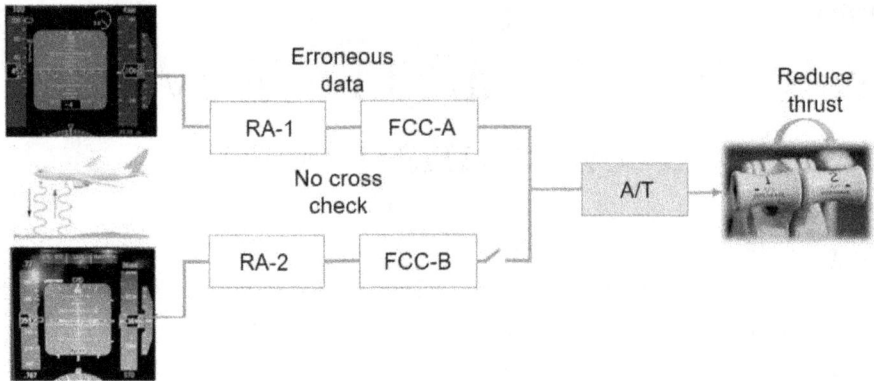

Figure 1.
Thrust reduction by autothrottle (A/T) using erroneous radio altimeter (RA) data [2].

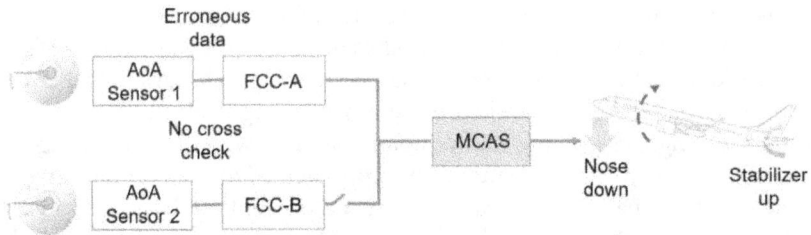

Figure 2.
Aircraft nose down by Maneuvering Characteristics Augmentation System (MCAS) using wrong angle of attack (AoA) data [2].

systems should be evaluated similarly to pilot error, focusing on its decision-making processes, ability to handle unexpected scenarios, and adherence to safety protocols, as well as identifying and mitigating potential "automation errors" that could arise from design flaws, misinterpretation of data, or unforeseen interactions.

In both systems, the Flight Control Computer (FCC-A or FCC-B) processes input data independently without cross-checking the data from the other line. This lack of cross-verification allows erroneous data from one input (e.g., RA-1 or AoA Sensor 1) to be used, resulting in incorrect actions. In the first system, the Radio Altimeter (RA-1) sends erroneous data to FCC-A, which directly influences the Autothrottle (A/T) system to reduce thrust. The absence of cross-checking between RA-1 and RA-2 prevents error detection, leading to improper actions. Similarly, in the second system, the Angle of Attack Sensor (AoA Sensor 1) provides incorrect data to FCC-A, triggering the Maneuvering Characteristics Augmentation System (MCAS) to push the aircraft nose down and the stabilizer up. The lack of cross-checking between AoA Sensor 1 and AoA Sensor 2 allows erroneous inputs to drive MCAS, leading to dangerous flight conditions.

The occurrence of both Lion Air and Ethiopian Airlines 737 MAX crashes resulting from inaccurate data from angle of attack (AoA) shows the criticality of AoA data activating MCAS. After these accidents, a comparator was added to compare the measurements from two AoA sensors as an error-proof design, as shown in **Figure 3**, in a way reducing the automation system error.

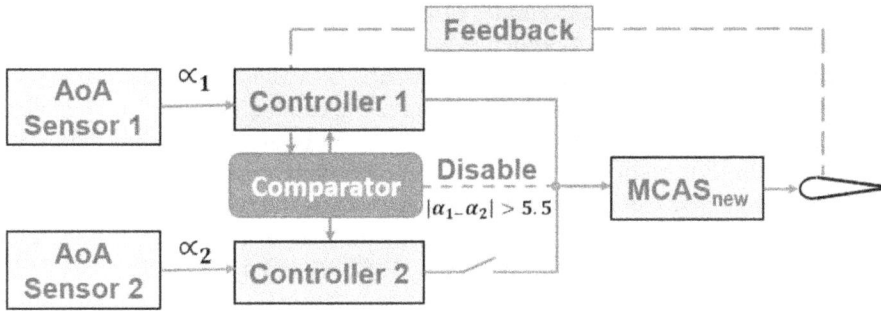

Figure 3.
Error-proof design of MCAS (MCAS$_{new}$) [2].

Figure 4.
The aircraft involved in accident [5].

Figure 5.
ATR-42 FQI (Left) and ATR-72 type FQI (Right).

The Air France crash in 2009 was another example of a similar computer mal-function [3]. Pruchnicki [4] said that it's quite easy to blame the pilots who died and claim that it has nothing to do with our poorly designed system. It gets frustrating because we keep having the same types of accidents.

One of the most illustrative examples of human-machine interaction errors is the ATR-72-202, shown in **Figure 4**, which experienced a dual-engine shutdown and ditched offshore near Palermo Airport in Italy [6].

Both ATR-42 and ATR-72 model aircraft have the same dimension and installation interface of fuel quantity indicator (FQI), as shown in **Figure 5**, which provides the pilots with the indication of the fuel weights in each wing tank. The only difference between the two FQIs, when installed on the aircraft, is the white lettering indicating the maximum fuel capacity for each wing fuel tank. This lettering, displaying "2500" for the ATR 72 type FQI and "2250" for the ATR 42 type, requires technicians to be highly attentive, as distinguishing between the two can easily lead to errors if not carefully checked. This means that an FQI for an ATR-42 airplane may be incorrectly installed on an ATR-72 and vice versa. After the accident, an airworthiness directive (AD) was issued to mandate a modification of the FQI installation to prevent incorrect installation. Such a development did not go beyond just solving a single problem. However, with the lesson learned here, aviation rules should have been developed so that similar parts for different model aircraft cannot be erroneously installed in place of each other, leading to incidents or accidents.

The final report on the crash [7] highlights that during a flight the day before the incident, the captain, who was also flying during the accident, identified an issue with the fuel quantity indicator (FQI) on the aircraft dashboard and reported it. Later that evening, the FQI was replaced with a unit intended for an ATR 42, a different aircraft model. This error occurred because the correct FQI was not located due to the part number being entered into the database in an incompatible format. Additionally, the inventory system incorrectly indicated that the ATR 42 part was compatible with both aircraft models.

Liu et al. [8] identified eight categories of human errors in human-machine interface design, which are misperception, memory lapse, carelessness, improper decision-making, improper operation, improper interface layout design, improper icon and text display design, and improper prompt feedback. They also state that current researchers mainly focus on feasibility, fabrication, and verification. So, they do not care enough about the human factor of users.

Latorella and Prabhu [9] reviewed human error in aviation maintenance and inspection to understand human error in aviation. Airbus [10] states that "Simple reactions like asking people to be "more careful" very rarely bring any improvement."

With the current safety management approach, human error is generally the root cause of safety investigations conducted by individual organizations. The most frequent corrective measure is to retrain people involved because human error is considered as the primary issue. However, this approach does not entirely solve the problems, and the errors reoccur. For example, emergency escape slides on Airbus 320 Family aircraft are error-prone and are frequently deployed inadvertently during maintenance in the region. Following numerous complaints from operators, Airbus redesigned the safety pin mechanism to reduce inadvertent deployments. The new design includes both an aural warning and lighting that activate when hands come close to the pin, effectively creating an error-proof system. As a result, such inadvertent deployments have significantly decreased. Another example is that an aircraft cockpit has two very close switches, and their names are written in a similar text and display design, which leads the user to select an incorrect button. As per traditional safety investigations into these situations, raising awareness through training, videos, and warnings for such conditions will only reduce human errors, not eliminate them.

When errors are investigated with the total system approach (TSA), many are classified as aircraft design and configuration or organizational factors rather than human errors. The total system approach (TSA), which will enable all relevant aviation stakeholders to be included in the research cycle not only for accidents but also for incidents issued as occurrence reports in 72 hours, has not yet been fully adopted. TEASAS [11] explained the TSA concept in aviation.

Although the need for a platform where safety issues and solutions for all occurrence reports are gathered and discussed in a shared data repository has come to the fore on different occasions, an effective solution for all stakeholders involving all stakeholders has not been in force in this regard yet. As a result, every organization should take care of itself on similar issues. Therefore, it is often challenging to produce effective solutions via safety investigations. Safety has no borders. For safety, we are all in the same boat, and we should work together to generate synergy.

2. Human factors integration

Aircraft are designed as per the regulations to ensure safety and airworthiness. The rules are published by major aviation authorities, such as the Federal Aviation Administration (FAA) and the European Union Aviation Safety Agency (EASA). The regulations are called certification requirements for commercial aircraft and originated from the International Civil Aviation Organization (ICAO) Annex 8 Airworthiness of Aircraft. When an aircraft is designed to comply with these rules, a type certificate is issued by the relevant authorities certifying that the aircraft can be manufactured accordingly. Additionally, an airworthiness certificate is issued for each aircraft manufactured, declaring that the particular aircraft is ready for flight.

Requirements for detailed design and construction provide a reasonable assurance that all airplane parts will function reliably and effectively [12]. When considering the safety and reliability of an aircraft, not only are the parts assembled but also the effects of those who operate and maintain the aircraft should be considered. In addition to the airplane itself, human reliability should be considered during the design phase, as reactive detection and resolution of human factor issues are very expensive and challenging compared to proactive human factors integration in operation [13].

Since the requirements do not cover the human reliability part, systems are designed without this consideration. As a total system approach to aircraft safety, the reliability of humans involved in the system in the life cycle should be included in the calculation, and the systems should be designed accordingly.

Regarding human factors, safety regulations are available for licensing pilots and technicians. These rules generally relate to the education, training, and experience required to obtain and maintain a license. However, these regulations do not guarantee that the licensed personnel does not make any mistakes.

When considering human factors in aircraft design from the perspective of pilots and technicians, the requirements for pilots are one step further. Although there are some rules about the cockpit design, many accidents are attributed to the pilots, even though they are design-related since the rules about the automation systems are insufficient [14]. Sherry et al. [15] studied designing user interfaces for the cockpit against five common cockpit automation design errors: (1) input devices that require reformulation of the mission task into subtasks or alternative representations, (2/3/4) absence of visual cues for access, format and insert (5) reformulation of display feedback for mission task verification and monitoring.

It is stated in the ICAO annexes booklet [12] that the configuration of the flight crew compartment must be such that the chances of improper control operation due to confusion, tiredness, or interference are as low as possible. It should allow a sufficiently clear, extensive, and undistorted field of vision for the safe operation of the airplane. Similar requirements for technicians should be added to the rules first by the ICAO and then by the authorities. Existing rules force the concepts of critical maintenance tasks (CMTs) and identical maintenance tasks (IMTs) for error capturing, placing responsibility on technicians rather than design. CMT tasks need to be independently checked by another mechanic, while IMT tasks need to be performed by different mechanics to reduce the risk of repeating the same error on both critical systems. When we examine the data from the Safety Management System (SMS) results, it is clear that there is a significant gap for safety improvements, which can be done with simple design changes. Although it is possible to reduce this gap only slightly with training, announcements, information, and similar applications, some improvements need to be made during the aircraft design, as the same types of events are repeated by different people. For this to become a standard, it is essential to review the aviation rules on these issues first.

3. Evaluation of human reliability

Massaiu and Paltrinieri [16] defined human reliability as "the probability of humans conducting specific tasks with satisfactory performance." The tasks for aircraft maintenance may be related to ensuring the continuing airworthiness of an aircraft or aircraft part, including cleaning, servicing, removal/installation, activation/deactivation, test, adjustment, inspection, check, operational/functional test, replacement, overhaul, defect rectification, the embodiment of modifications, compliance with airworthiness directives, and repair. The pilots' tasks may be related to flying the aircraft, including monitoring all systems, preparing a flight plan, communicating with air traffic control or ground personnel, consideration of aircraft performance, altitude, and weather conditions. Alogla and Alruqi [17] gave the classification of human error. Humans are everywhere from pre-design, design development, and manufacturing to operating a system or machine. The human errors mentioned in this study cover the process of the use of the designed product. Human factors in the design of the product should be examined under machine reliability, which is beyond the scope of this study. Of course, all machines are made by people and have reliability. Although robots can make machines, they are also machines designed by humans, so the line has to be drawn somewhere, where the best distinction is in the form of design and use process. The designer is actually the actor of the machine. The maintenance and use of the machine are continued in line with the directives foreseen by this actor. In incident and accident investigations, if the event is related to the pilot or technician, it is considered a human error, if it is related to design and production, it is considered a machine error. In fact, there is no problem with this distinction in practice. The main issue is that when handling errors, a machine-related error is usually logged as a user error, as with the first 737 MAX crash. Had it not been for the second 737 MAX crash, it would likely have gone on record as a human error. The 737NG Amsterdam accident caused by the altimeter in 2009 was another example of a human error, although it was a machine error because altimeters did not have cross-checks during wrong incoming data. It came to the fore after the 737MAX accidents, where this accident was misjudged [2].

Munot [18] conducted a study on Human Error or Design Error and stated that "In many fields 60–80% of all accidents are due to human error. These assumptions are misleading because they think that a person should have taken or not taken a possible action, but the real source of the error is the design rather than the human."

Senders and Moray [19] defined human error as "something having been done that was not intended by the actor; not desired by a set of rules or an external observer; or that led the task or system outside its acceptable limits." Designers are actors who plan and develop machines. Machines generally consist of two parts, hardware and software. Although the software and hardware are designed and developed by humans, they should be evaluated under machine reliability. Whether the machine reliability is high or low directly depends on the designer, who is the actor. In the traditional approach, redundancy and similar techniques are used to ensure machine reliability. The shortcoming here is that when designers calculate the reliability of the machines, they often ignore the possibility of users making mistakes.

Alogla and Alruqi [17] made a study about Aircraft Assembly Snags: Human Errors or Lack of Production Design? and stated that "One major industry that heavily relies on human work is the aerospace industry." They also expressed that "Designing prevention systems will lead to a decrease in the proportion of workers' errors."

In terms of machine reliability itself, type certification requirements are available. In line with the developments in automation and similar technologies, these rules need to be updated continuously. Sometimes, because technological advances are not followed adequately, the rules lag behind them, which leads to undesirable results. During the design and development of the aircraft as a machine, designers may also make mistakes. Since these types of errors are in the process of design or manufacturing phase of a part before it is ready for use, these errors are considered within the scope of machine reliability.

Hawkins [20] presented the SHEL model driven from the initial letters of four components, which are Software, Hardware, Environment, and Liveware, as a valuable tool for analyzing human factors. In aviation, liveware is the interface between the users, pilot or mechanic, and machine. In this model, hardware- and software-related issues are evaluated within the scope of machine reliability.

Sgobba et al. [21] emphasized the need for improvement in machine and human interface to reduce human errors. They concluded that "Unfortunately, in many projects, human error is not considered during RAM (Reliability, Availability, and Maintainability) analysis based on the assumption of no human error. In fact, human error can be identified in the early asset life cycle design during FMEA as a cause of failure mode."

The probability of a human error P(HE), which is human unreliability, can be written as follows,

$$P(HE) = \frac{Number\ of errors}{Number\ of\ times\ performing\ a\ task} \tag{1}$$

Since the sum of reliability and unreliability is one at any time, human reliability R_{Human} can be found below,

$$R_{Human} = 1 - P(HE) \tag{2}$$

The greater number of error opportunities are, the more errors occur, which means higher human unreliability. So, in order for the human reliability to increase, error opportunities due to machine and human interaction should be eliminated as much as possible. Eliminating the error opportunities during the design significantly reduces accidents and incidents. Blaming people for mistakes made, even though the designs are not error-proof, is equivalent to not seeing the root cause of the errors.

They are human and to err is human. As long as aircraft are not maintenance-free and self-driving, technicians and pilots are part of the total system.

Corteney et al. [22] explored a new approach to identifying a potential human error in helicopter maintenance to find out where error-proofing features or other mitigations are most needed to support the maintenance engineer during critical maintenance tasks. They stated that a maintenance task for continued airworthiness should occur at a specific interval. The MSG-3 (Maintenance Steering Group) approach does not require a comparison of the planned maintenance task to the limitations and realities of human performance, as well as known types of human errors. The main innovation brought about by this study is the inclusion of human reliability in the MSG-3 logic using the total system approach (TSA). TSA is used to integrate and optimize all the activities of each segment in the life cycle [23]. TSA is a system that includes both machine and human elements that perform a function. The approach can evaluate aircraft safety, as shown in **Figure 6**.

When calculating the reliability, R, of the aircraft systems with a total system approach, in addition to the aircraft system or part, which may be called a machine too, human reliability should be considered, as shown in **Figure 7**. A machine can have either or both software and hardware.

Four parts of the system are connected in series because if anyone fails, the whole system fails. For the total system to be reliable, all of them must be reliable. So, R_{Total}, the total system reliability of such a system, can be calculated as below.

$$R_{Total} = R_{Hardware}(t)R_{Software}(t)R_{Technician}(t)R_{Pilot}(t) \qquad (3)$$

Since the reliability values are between 0 and 1, R_{Total} is less than the lowest reliability. This means that the reliability of a series system is always lower than

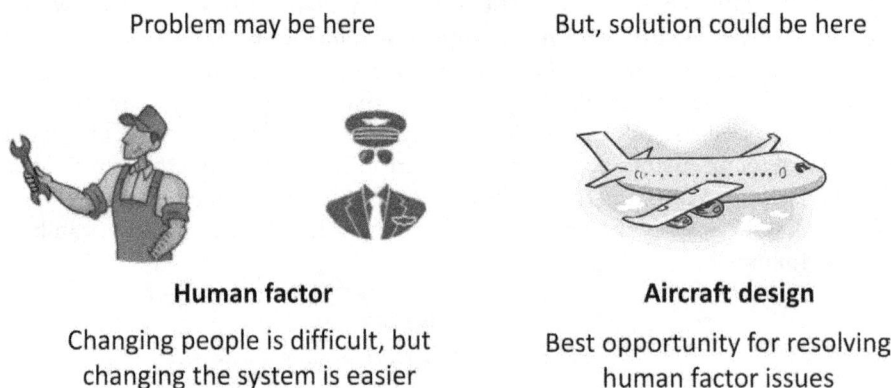

Problem may be here · But, solution could be here

Human factor

Changing people is difficult, but changing the system is easier

Aircraft design

Best opportunity for resolving human factor issues

Figure 6.
Total system approach to aircraft safety.

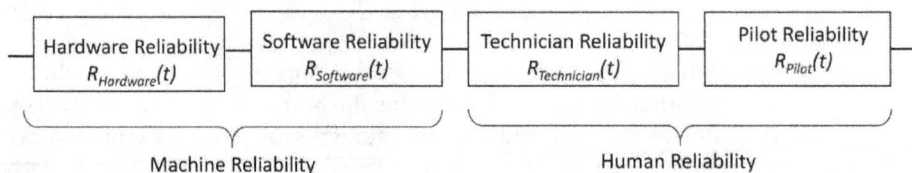

| Hardware Reliability $R_{Hardware}(t)$ | Software Reliability $R_{Software}(t)$ | Technician Reliability $R_{Technician}(t)$ | Pilot Reliability $R_{Pilot}(t)$ |

Machine Reliability · Human Reliability

Figure 7.
Graphical representation of total system.

that of individual part reliability, and the total system reliability decreases faster over time.

The reliability function can be written as below [24],

$$R(t) = e^{-\int_0^t \lambda(t)dt} \tag{4}$$

where R(t) is the reliability at time t and $\lambda(t)$ is the failure rate. Using Eqs. (3) and (4), R_{Total} can be found below.

$$R_{Total}(t) = e^{-\int_0^t \lambda_H(t)dt} e^{-\int_0^t \lambda_S(t)dt} e^{-\int_0^t \lambda_T(t)dt} e^{-\int_0^t \lambda_P(t)dt} \tag{5}$$

where $\lambda_H(t)$ is the failure rate of the hardware at time t, $\lambda_S(t)$ is the failure rate of the software at time t, $\lambda_T(t)$ is the error rate of the technician at time t, and $\lambda_P(t)$ is the error rate of the pilot at time t. Since there are generally no data on error rates during design, estimation methods are used. Shirali et al. [25] outlined human error rate estimation methods.

Givi et al. [26] conducted a study about modeling worker reliability with learning and fatigue. Giuntini [27] gave an estimation of human error rate behaviors using a bathtub curve representing three phases, which are the learning phase, stabilized error phase, and fatigue phase. Depending on the experience, the error rate may decrease with time first during the learning phase. The error rate stabilizes and reaches a constant value as the user gains knowledge. In the fatigue phase, the error rate increases over time. Luis et al. [28] surveyed stress, pressure, and fatigue on aircraft maintenance personnel. The survey results raised serious concerns that exposure to stress, fatigue, and pressure could lead to maintenance errors.

As the learning phase starts again with newcomers and fatigue occurs due to long work hours, irregular work shifts, time pressure, and similar contributing factors common in the aviation community, three phases are experienced together generally in an organization during the same period. So, in the study, it is reasonable to use average error rates for technicians and pilots, as seen in the bathtub curve in **Figure 8**.

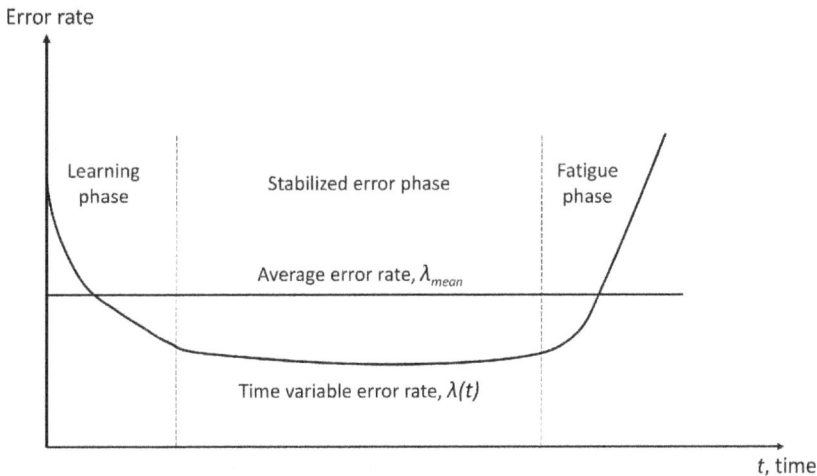

Figure 8.
Bathtub curve for error rate.

It is also possible to use the Weibull distribution when it is desired to model all three regions separately, instead of the mean [24]. There are other techniques to estimate human reliability as well. Catelani et al. [29] proposed an innovative procedure that uses Artificial Intelligence (AI) and Fuzzy logic to assess human reliability in railway engineering. Huang [30] studied human reliability analysis in aviation maintenance by a Bayesian Network approach. Kontogiannis [31] emphasized the new approaches to data collection on human performance for predicting human errors. Swain and Guttmann [32] developed the human error rate prediction technique. Askren [33] studied quantifying human performance reliability using the two-parameter Weibull function.

With this assumption, error rates for the technician and pilot can be written as below.

$$\lambda_T(t) = \lambda_T \text{ and } \lambda_P(t) = \lambda_P \tag{6}$$

Where λ_T is the average technician error rate and λ_P is the average pilot error rate. Using Eqs. (5) and (6), R_{Total} can be written as below.

$$R_{Total}(t) = e^{-\int_0^t \lambda_H(t)dt} e^{-\int_0^t \lambda_S(t)dt} e^{-(\lambda_T+\lambda_P)t} \tag{7}$$

From this equation, total system reliability can be found, according to the design of the system. If the system is designed as error-proof for only technicians or pilots, the total reliability function can be written in Eqs. (8) and (9), respectively.

$$R_{Total}(t) = e^{-\int_0^t \lambda_H(t)dt} e^{-\int_0^t \lambda_S(t)dt} e^{-\lambda_P t} \tag{8}$$

$$R_{Total}(t) = e^{-\int_0^t \lambda_H(t)dt} e^{-\int_0^t \lambda_S(t)dt} e^{-\lambda_T t} \tag{9}$$

If the system is designed as error-proof for both technicians and pilots, $\lambda_T = \lambda_P = 0$. In this case, the total reliability function is given as below,

$$R_{Total}(t) = e^{-\int_0^t \lambda_H(t)dt} e^{-\int_0^t \lambda_S(t)dt} \tag{10}$$

If the machine has either hardware or software in a system designed as error-proof, the total reliability function can be written for both situations, respectively,

$$R_{Total}(t) = e^{-\int_0^t \lambda_H(t)dt} \tag{11}$$

$$R_{Total}(t) = e^{-\int_0^t \lambda_S(t)dt} \tag{12}$$

If an automation system replaces the human action, system reliability depends only on the machine, which means either or both software and hardware.

Aircraft safety-critical systems are designed so that no more than one accident occurs in 10^9 flight hours (FH). So, the expected reliability of such a system, $R_E(T)$, can be found below,

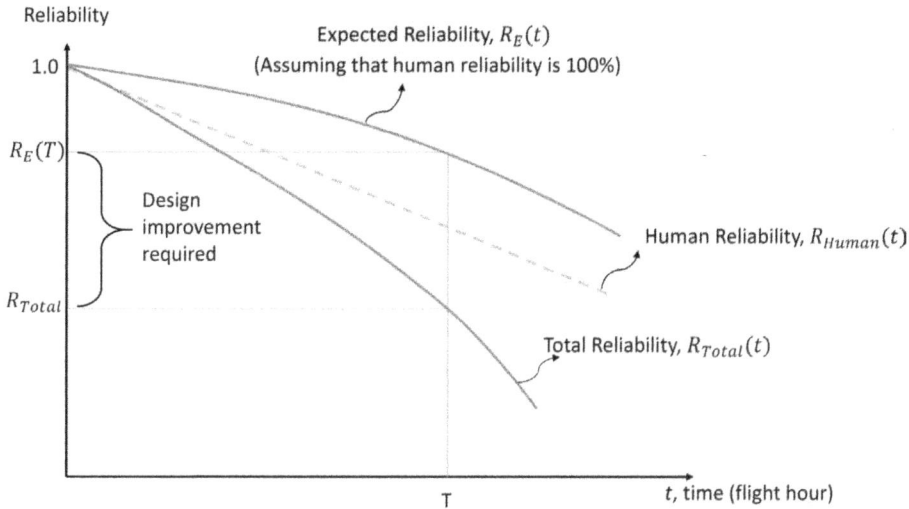

Figure 9.
Design improvement requirement due to human reliability.

$$R_E\big(T = 10^9\ FH\big) = 1 - \frac{1}{10^9} \tag{13}$$

where $\frac{1}{10^9}$ is the unreliability of the system. The sum of reliability and unreliability at any time is zero. That's why it was subtracted from 1 to find the reliability of the system. The change in total reliability due to human reliability is shown in **Figure 9**.

In the traditional approach, reliability calculation is done assuming that human reliability is 100%. So, expected reliability can be replaced with machine reliability since it is given based on machine reliability itself. As seen from the graph, although the design is perfect in terms of functioning, if pilots and technicians are prone to making mistakes in this system, the total system reliability decreases. Therefore, design improvement in terms of error-proof is needed to reach the expected reliability required by the authorities.

4. Error-proof design methods and principles

Error-proof design is to design the parts or systems to prevent or at least reduce as much as possible human error by increasing error resistance. There are three primary levels of error-proofing: awareness, detection, and prevention, as seen from the chart in **Figure 10**. The awareness level generally involves training, warnings, and work instructions with the lowest error resistance. The detection level includes independent or double inspections and visual aids, such as color codes, signs, and labels, which help determine human errors before releasing them to the service. The prevention level, which utilizes methods, such as Poka-Yoke and Murphy, is the most comprehensive error-proofing level, with the highest error resistance.

It is required to consider the effect of the error on this subject to predict which level of error-proofing might be sufficient during the design. The more impact of the

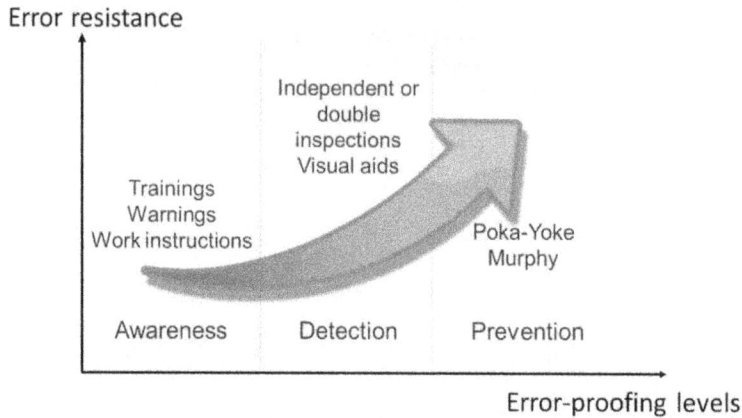

Figure 10.
Effect of error-proofing levels on error resistance.

foreseeable errors increases from minor to catastrophic, the higher the error-proofing level should be. Courteney et al. [22] gave recommendations about error-proofing requirements depending on the effect of the predictable errors on safety. For example, for catastrophic consequences of predictable mistakes, the error should be prevented by design or the criticality of the system reduced to a lower level.

Dunn [34] states that "with the traditional engineering approach to dealing with maintenance error, most engineers tend to think along two lines – either discipline/counsel/train the individual(s) involved, and/or write a new procedure/work instruction to make sure that it doesn't happen again. Unfortunately, recent research and experience by Behavioral Psychologists indicate that neither of these approaches is likely to be successful in eliminating maintenance error." Although the third one is the most fault-tolerant system, it is often almost impossible for the aircraft operators to design themselves. The improvements are mostly done at level 1, awareness, and sometimes at level 2, detection. Therefore, design-based advances against errors after use are generally ignored. It is not only cheaper and easier to make such improvements during design.

Capt. Edward A. Murphy, an aircraft designer working on Air Force Project MX981 in the 1940s, found that a transducer was wired wrongly by a technician. Then he said, "If there is any way to do it wrong, he'll find it." And with this statement, the foundations of Murphy's law were laid. According to Murphy's law, "if anything can go wrong, it will" (murphy-laws.com). This may be a pessimistic idea, but this is the case in reality. Therefore, the more you reduce the chance of something going wrong, the safer it will be. If a part or component is designed without thinking of Murphy's concept, the design is called no murphy design. **Figure 11** shows two no murphy design examples of parts on an aircraft type with no murphy.

Since the shapes of the hose nozzles in **Figure 11** are the same, they can be confused and installed in place of each other during assembly by people who have little or no experience in the subject. Although experienced persons have a decreased error rate, they are nevertheless involved in similar mistakes. In such cases, providing information and training and adding warnings to the documents do not prevent the recurrence of the mistakes. Manufacturers consider it is sufficient to add notifications

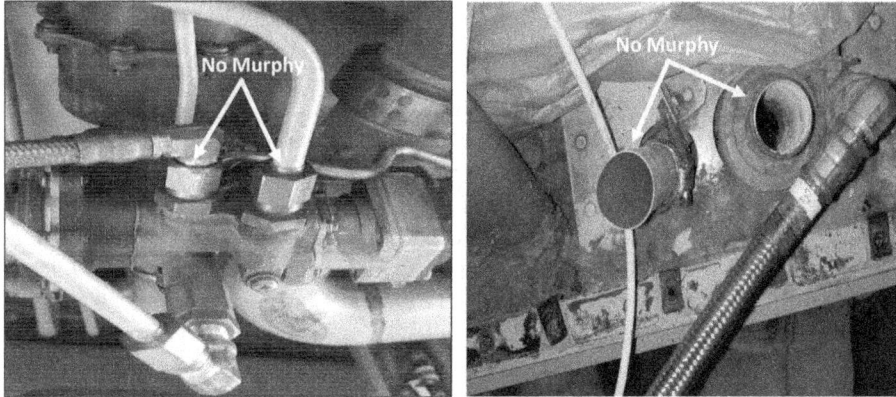

Figure 11.
No murphy design examples.

to maintenance documents to avoid mistakes. However, these improvements, which remain at the awareness level, cannot prevent errors.

The concept of Poka-Yoke includes changes made during design to minimize the possibility of error by adopting the idea that humans are not 100% reliable. The Poka-Yoke concept, developed by Shigeo Shingo in the 1960s, is to design a system so that mistakes or error opportunities are detected and corrected at the source [35]. Poka and Yoke are Japanese words that mean "to avoid" and "inadvertent mistake," respectively [36]. Bos [36] gave the details about how Poka-Yoke is implemented.

Error-proof design is not a new idea. The concept has been applied to different parts or components in aircraft. However, there is no propagation to use the concept as a standard similar to the fail-safe design mandated by aviation rules. Therefore, there are still errors that can be avoided with elementary design changes, and some of them lead to severe consequences, up to fatal accidents. This kind of design change should be made during the initial design development rather than after use. Although it is more costly than the design phase, continuous improvement should be made by applying modifications for unforeseen situations during the design in light of data obtained through occurrence reports.

ICAO [37] Annex 8, Airworthiness of Aircraft, Paragraph C.8 states that "The design and construction of the airplane shall conform to damage tolerance, safe-life or failsafe principles and shall be such as to ensure that the probability of catastrophic failure during the operational life is extremely remote." Since the fail-safe requirement is available in ICAO, the FAA and EASA, major aviation authorities, have rules about it. Similarly, error-proof design should be included in this paragraph to be a generally accepted standard and to be applied by everyone.

5. Designing error-proof aircraft

The safety management system (SMS) concept was developed by the International Civil Aviation Organization (ICAO) in 2006 to increase safety by taking a systematic approach to human errors [38]. The SMS approach has supported the industry in enhancing human performance and reducing the number of human errors. However, it has not been possible to prevent recurring mistakes from time to time, or it is not easy to find a solution because the system is not designed to be error-proof. Since there are no binding rules for designing aircraft considering human factors, especially for maintenance, aircraft designers sometimes do not support humans by considering human factors. FAA provides guidance [39] on installing flight control systems in the cockpit from a human reliability perspective. Sun et al. [40] emphasized the importance of error-proof design of controls in aircraft cockpits to reduce operational errors. But, these rules are mostly related to ergonomics and they do not address flight controls or other controls not located in the flight deck.

During the design, failure modes and effects analysis (FMEA), fault tree analysis (FTA), and similar analysis techniques are used to understand whether the reliability of aircraft systems and components is at the desired level and MSG-3 logic is used to develop maintenance programs to keep the aircraft's reliability at the desired level. However, these analyses focus on machine faults, not human errors. Therefore, it is necessary to make sure that aircraft is designed to be error-proof by considering the possibility of errors.

Since it is not feasible for every part of the aircraft to be error-proof, the effects of errors should be identified first. In the study, the error effect categories and required error-proofing techniques are developed, inspired by MSG-3 logic (ATA, 2001) and Reliability Centered Maintenance (RCM) [41].

Depending on the evidence and impact of error, error effect categories are defined in the diagram as shown in **Figure 12**. In the provided chart, the numbering system (1–9) is used to logically structure the decision-making process for categorizing errors based on their evidence and impact. (1): Is error evident to technician or pilot? This is the first decision point, determining if the error is apparent to the user (technician or pilot). (2) and (3) are secondary decision points that follow the first. (2): If the error is evident, does it affect operating safety? (3): If the error is hidden, does it affect operating safety? Then, the decision point (4) applies to evident errors that do not affect safety. (4): Does error affect operating capability?

Based on the evidence and impact of error, error effect categories are defined as (5) Evident Safety, (6) Evident Operational, (7) Economic, (8) Hidden Safety, and (9) Hidden Economic, as shown in **Figure 12**.

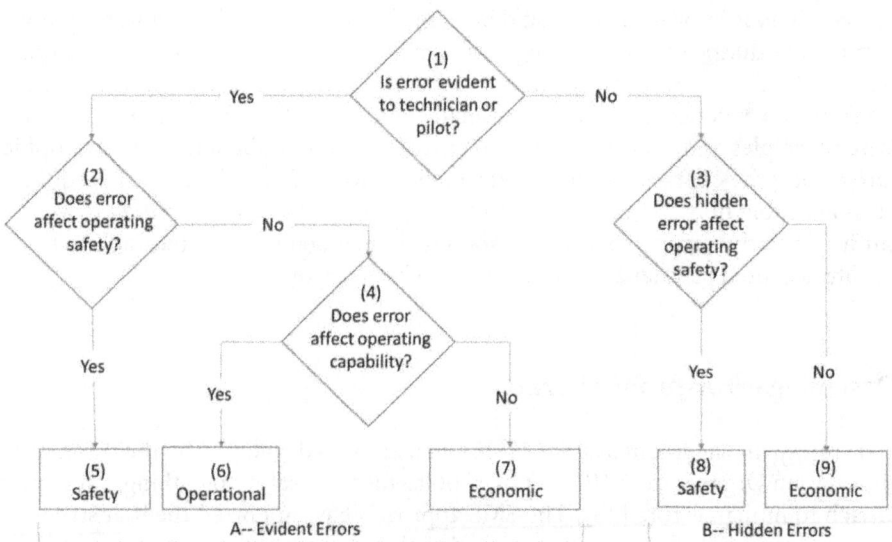

Figure 12.
Error effect categories.

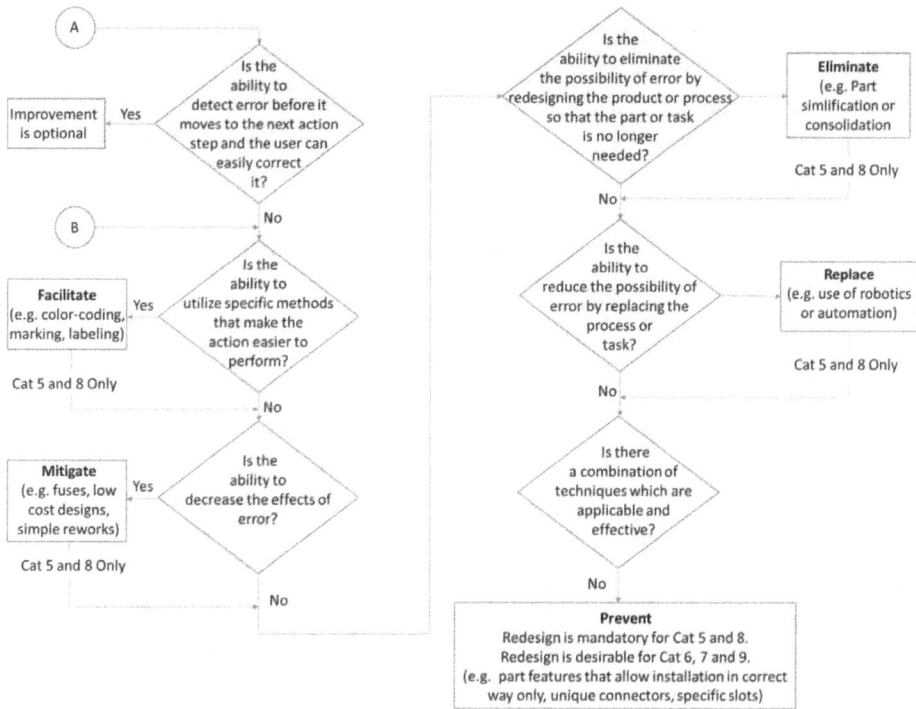

Figure 13.
Error-proofing technique selection and assignment.

NPDSolutions [42] and McBride [43] explained the error-proofing methods with the examples of the principles. NPDSolutions stated that "mistake-proofing should be considered during the development of a new product to maximize opportunities to mistake-proof through the design of the product and the process. Once the product is designed and the process is selected, error-proofing opportunities are limited." Error-proofing techniques can be ordered as (1) Facilitation, (2) Mitigation, (3) Elimination, (4) Replacement, and (5) Prevention with the degree of difficulty and effectiveness.

Depending on the error effect category, all the following questions must be answered to define the error-proofing techniques that should be assigned, as shown in **Figure 13**.

- Is the ability to detect error before it moves to the next action step, and the user can easily correct it?

- Is the ability to utilize specific methods that make the action easier to perform?

- Is the ability to decrease the effects of error?

- Is the ability to eliminate the possibility of error by redesigning the product or process so that the part or task is no longer needed?

- Is the ability to reduce the possibility of error by replacing the process or task?

- Is there a combination of techniques that are applicable and effective?

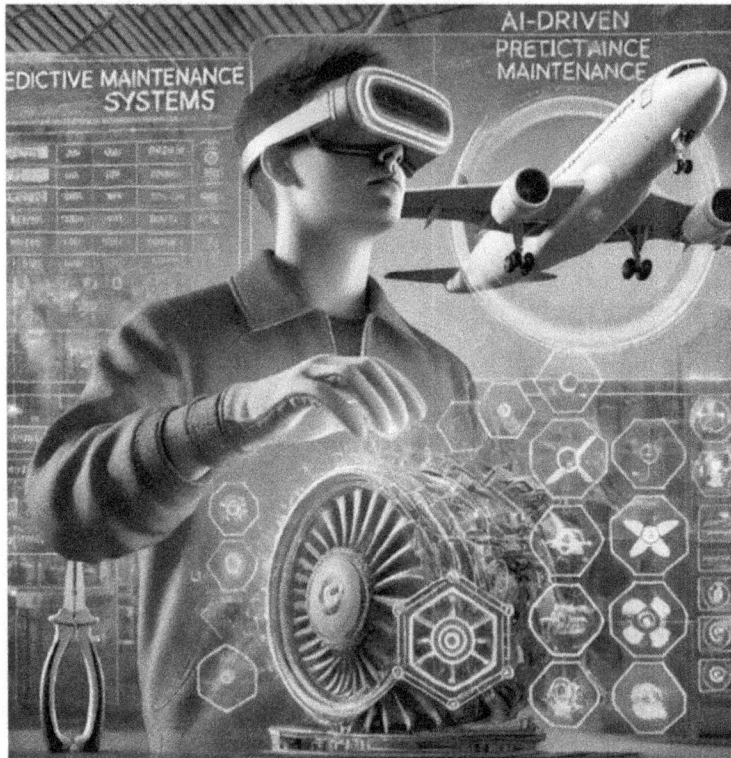

Figure 14.
Integration of AI-powered predictive maintenance and augmented reality (AR).

As seen from the figure, all the questions must be asked for the errors in categories (Cat) 5 and 8 because they are directly related to safety. If the error is in the evident category and can be detected by the pilot or technician, then an error-proof design is not required. For the other categories, i.e. categories (6), (7), and (9), in the case where a solution is found for any question asked, there is no need to continue to the next question. If the answer to the last question is no for Cat 5 and 8, then redesign is mandatory. For the other categories, redesign is desirable. For example, if the questions are asked for 737 MAX MCAS, which led to two deadly accidents, it is seen that redesign is mandatory.

Looking to the future, advances, such as AI-powered predictive maintenance systems and augmented reality for technician and pilot training, can further reduce human error. **Figure 14**, generated using AI, illustrates how these technologies provide real-time feedback and predictive diagnostics, enhancing safety and efficiency in aircraft operations and maintenance. Integrating these innovations and error prevention principles into the aviation industry can proactively address potential risks and optimize performance.

6. Moving toward proactive error-proofing regulatory standards

The final report of the ATR 72-200 accident [7] states that "At present, there are no general regulations that oblige manufacturers to provide installation modifications for components with the same functions and ostensibly similar from a constructive point of view, but with different performance, which may be installed on various types/versions of aircraft belonging to the same family (e.g., Boeing B737-200, -400, -800; Airbus A319, A320, A321; ATR 42 and 72, etc.). Consider the possibility of carrying out studies aimed at defining guidelines and/or issuing regulatory requirements concerning the provision of suitable installation modifications on the aircraft or the component itself, to avoid cases where components with the same functions and ostensibly similar but with different performances could be installed in error." As highlighted in the safety recommendation from the accident report, there is no rule or requirement for an error-proofing system in current regulations. This emphasizes the urgent need for clear guidelines and regulatory standards to prevent incorrect installations across different aircraft types. If such a system had been mandated, the two fuel indicators could not have been designed in a way that allowed the wrong installation, and this accident would not have occurred. Incorporating such requirements into aviation standards would ensure that manufacturers provide suitable installation modifications, thereby mitigating the risk of design-related errors and enhancing overall operational safety.

Certification requirements must be developed to effectively implement proactive error prevention standards. Current certification processes are primarily focused on demonstrating compliance with existing safety requirements. A shift toward proactive standards would involve updated certification tests and verification procedures specifically for error-proofing capabilities. For instance, certification could require testing for specific error-proofing mechanisms in simulated environments that mimic potential operational challenges.

Integrating error-proofing criteria in certifications would also address design oversights, such as reliance on single-sensor data. Updated certifications should require manufacturers to demonstrate robust error-proofing across all phases of an aircraft's life cycle, including design, production, maintenance, and operation. This approach emphasizes safety from the outset, rather than relying on retrofitting solutions after issues arise. Regulatory standards should require that aircraft be designed using Error Prevention Technique Selection and Assignment in compliance with Error Effect Categories, as explained in the previous section.

However, robust error-proofing cannot be achieved without transparent data sharing across the aviation industry. The lack of such sharing has already led to preventable tragedies. For instance, if a comprehensive database had existed to collect and analyze comments after the Lion Air accident, the second accident could have been avoided [12]. Unfortunately, economic concerns often hinder access to such data, resulting in more lives lost. The two Boeing 737 MAX crashes in 2018 and 2019, 5 months apart, claimed 346 lives and severely damaged trust in automation systems. These incidents raised questions about certification standards and led to the global grounding of an entire aircraft type for almost 2 years due to a design flaw, highlighting the importance of lessons learned from the perspective of the total system approach [2].

Initiatives like Data4Safety (D4S) [44] in Europe demonstrate the potential of collaborative platforms to identify risks, refine safety measures, and enhance system resilience through big data analytics and benchmarking. However, safety has no borders, and limiting such efforts to specific regions leads to gaps in comprehensively addressing widespread risks. ICAO, a leading global aviation governance authority with 193 member countries representing nearly the entire global aviation community [45], is well positioned to take responsibility for establishing a centralized data warehouse where safety data, along with recommendations and comments, can be shared transparently. Establishing such a global safety data platform, rooted in just culture, robust data protection, and global collaboration, enables manufacturers to develop resilient, error-proof systems, reduce risks, enhance safety, and foster a unified aviation ecosystem.

7. Conclusions and recommendations

Contrary to traditional thought, machines rather than humans cause many incidents and accidents. Humans do not knowingly make mistakes, and when it does, it is considered a violation. The critical thing here is to understand why errors are made and how to prevent them. Here is where the error-proof design comes in. To determine when an error-proof design is required, reliability calculations need to be done, considering human reliability to achieve the expected reliability with the total system approach. Design improvement is required when reliability levels are below the expectation level due to error opportunities, and error-proof techniques must be used to increase human reliability. In eliminating design-related errors, error resistance should be increased during the design phase rather than waiting for a solution from information and education against these error opportunities. For such a change to become active, starting from the authorities and covering all the stakeholders throughout the chain, ICAO should be encouraged to adopt the total system approach within the safety management system. Then, aircraft-type certification requirements must be updated accordingly. Similar to design deficiencies, which cause failures, the authorities should publish airworthiness directives to make error-proof design improvement mandatory when necessary. Otherwise, operators cannot produce effective solutions on their own. A centralized platform should be established for authorities, manufacturers, operators, and MROs to share and analyze safety issues, ensuring error-proof designs are guided by real-life data and examples. Transparency and collaboration among aviation stakeholders are important to advance safety standards and foster innovation across the industry.

Moreover, designing error-proof aircraft not only enhances safety but also provides a competitive advantage in the market. Airlines are increasingly drawn to manufacturers that prioritize reliability and operational efficiency, as these factors directly impact their profitability and customer satisfaction. By integrating error-proofing principles into aircraft design, manufacturers can differentiate their products, attract more buyers, and position themselves as industry leaders committed to innovation and safety. This approach can also open new marketing opportunities, as error-proof designs reduce downtime, maintenance costs, and the risk of operational disruptions, making them highly appealing to prospective buyers.

Adopting the error-proofing logic for technique selection and assignment based on error effect categories in aircraft design, as in MSG-3, will enhance human reliability and enable a more accurate determination of the root causes of events.

Acronyms and abbreviations

AD	airworthiness directive
AI	artificial intelligence
AR	augmented reality

AoA	angle of attack
A/T	Autothrottle
Cat	category
CMT	critical maintenance task
D4S	Data4Safety
EASA	European Union Aviation Safety Agency
FAA	Federal Aviation Administration
FCC	flight control computer
FQI	fuel quantity indicator
FMEA	failure modes and effects analysis
FTA	fault tree analysis
HE	human error
IMT	Identical maintenance tasks
ICAO	International Civil Aviation Organization
MCAS	Maneuvering Characteristics Augmentation System
MSG-3	maintenance steering group-3
MRO	maintenance, repair, and operations
RA	radio altimeter
RAM	reliability, availability, and maintainability
RCM	reliability centered maintenance
SMS	safety management system
TSA	total system approach

Author details

Seref Demirci
Saudia Technic, Jeddah, Saudia Arabia

*Address all correspondence to: sherefdemirci@gmail.com

IntechOpen

© 2025 The Author(s). Licensee IntechOpen. This chapter is distributed under the terms of the Creative Commons Attribution License (http://creativecommons.org/licenses/by/4.0), which permits unrestricted use, distribution, and reproduction in any medium, provided the original work is properly cited. [cc] BY

References

[1] Boeing. MEDA investigation process. 2008. Available from: https://www.boeing.com/commercial/aeromagazine/articles/qtr_2_07/article_03_2.html

[2] Demirci S. The requirements for automation systems based on Boeing 737 MAX crashes. In: Aircraft Engineering and Aerospace Technology. UK: Emerald Publishing Limited; 2021. DOI: 10.1108/AEAT-03-2021-0069

[3] Engelmann E. Designing to prevent failure: Poka Yoke.2012. Available from: https://geovoices.geonetric.com/2012/07/designing-to-prevent-failure-poka-yoke/

[4] Pruchnicki S. Turkish Airlines Crash Probe only Told Half the Story. NY, USA: New York Times; 2020

[5] Tuninter Flight 1153. Available from: https://en.wikipedia.org/wiki/Tuninter_Flight_1153

[6] ANSV. Safety recommendations about the accident occurred on 6 August 2005 offshore Palmero Airport (Sicily, Italy). 2005. Available from: https://skybrary.aero/sites/default/files/bookshelf/2435.pdf

[7] ANSV. Final report, accident involving ATR 72 Aircraft Registration TS-LBB ditching off the coast of Capo Gallo (Palermo - Sicily). 2005

[8] Liu X, Liu Z, Chen P, Xie Z, Lai B, Zhan B, et al. Human Reliability Evaluation Based on Objective and Subjective Comprehensive Method Used for Ergonomic Interface Design. Amsterdam, Netherlands: Elsevier; 2021

[9] Latorella KA, Prabhu PV. A review of human error in aviation maintenance and inspection. International Journal of Industrial Ergonomics. Amsterdam, Netherlands: Elsevier; 2000;**26**

[10] Airbus. Human Performance Error Management. Toulouse, France: Flight Operations Briefing Notes; 2005

[11] TEASAS. Total System Approach in Aviation. North Carolina, USA: McFarland; 2022. Available from: http://www.teasas.com/total-system-approach/

[12] ICAO. Annexes Booklet 1 to 18. Montreal, Canada: ICAO; 1974. Available from: https://www.icao.int/safety/airnavigation/nationalitymarks/annexes_booklet_en.pdf

[13] ICAO. Human Factors Training Manual, Doc 9683iAN/950. Montreal, Canada: ICAO; 1998

[14] Demirci S. The role of consensus standards on safety in the light of Boeing 737 MAX accidents. In: 11th Ankara International Aerospace Conference, AIAC-2021-040. Ankara, Turkiye. 2021

[15] Sherry L, Polson P, Feary M. Designing User-Interfaces for the Cockpit. Pennsylvania, USA: Society of Automotive Engineers, Inc.; 2001

[16] Massaiu S, Paltrinieri N. Human Reliability. Amsterdam, Netherlands: ScienceDirect; 2016. Available from: https://www.sciencedirect.com/topics/engineering/human-reliability

[17] Aloga AA, Alruqi M. Aircraft assembly snags: Human errors or lack of production design? Aerospace. 2021;**8**: 391. DOI: 10.3390/aerospace8120391

[18] Munot S. Human Error or Design Error?. San Francisco, California, USA: UX Planet; 2017. Available from: https://

uxplanet.org/human-error-or-design-error-23facef8f36e

[19] Senders JW, Moray NP. Human Error: Cause, Prediction, and Reduction. NJ, USA: Lawrence Erlbaum Associates; 1991. p. 25

[20] Hawkins FH. Human Factors in Flight. Aldershot: Ashgate; 1994

[21] Sgobba T, Kanki B, Clervoy JF, Sandal GM. Space Safety and Human Performance. Amsterdam, Netherlands: Elsevier; 2018

[22] Courteney H, Gill S, Carmichael S. Designing out human error. 2018. Available from: https://www.aerosociety.com/news/designing-out-human-error/

[23] Min H. Essentials of Supply Chain Management, the New Business Concepts and Applications. NJ, USA: FT Press; 2015

[24] Demirci S. Reliability and Maintainability Analysis with an Application to Aircraft Maintenance. İstanbul Technical University, Institute of Science and Technology; 1998

[25] Shirali GA, Hosseinzadeh T, Angali KA, Kalhori RN. Modifying a Method for Human Reliability Assessment Based on CREAM-BN: A Case Study in the Control Room of a Petrochemical Plant. Elsevier; 2019. DOI: 10.1016/j.mex.2019.02.008

[26] Givi ZS, Jaber MY, Neumann WP. Modelling worker reliability with learning and fatigue. Applied Mathematical Modelling. 2015;**39**(17): 5186-5199

[27] Giuntini RE. Mathematical Characterization of Human Reliability for Multi-Task System Operations. NJ, USA: IEEE; 2000

[28] Luis F, Santos FM, Melicio R. Stress, pressure and fatigue on aircraft maintenance personal. Vol. 12. No. 1. Napoli, Italy: International Review of Aerospace Engineering (IREASE); 2019; **12**(1)

[29] Catelani M, Ciani L, Guidi G, Patrizi G. Human error probability estimation for safety and diagnostic systems in railway engineering. Measurement: Sensors. Amsterdam, Netherlands: Elsevier; 2021;**18**

[30] Huang SP. Human reliability analysis in aviation maintenance by a Bayesian network approach. In: Conference: ICOSSAR2013. New York: Columbia University; 2013

[31] Kontogiannis T. Technical Evaluation Report. Chania, Greece: Department of Production Engineering and Management Technical University of Crete University Campus; 1999

[32] Swain AD, Guttmann HE. Handbook of human reliability analysis with emphasis on nuclear power plant applications, NUREG/CR-1278, USNRC. 1983

[33] Askren WB. Quantifying human performance reliability. 1971. Available from: https://files.eric.ed.gov/fulltext/ED079349.pdf

[34] Dunn S. Managing human error in maintenance. Maintenance & Asset Management. 2014;**20**(4)

[35] Robinson H. Using Poka-Yoke techniques for early defect detection. In: Sixth International Conference on Software Testing Analysis and Review. Florida, USA: Software Quality Engineering; 2012

[36] Bos PSC. What Is Poka-Yoke?. San Francisco, California, USA: Academia; 2014. Available from: https://www.

academia.edu/33564796/Poka_Yoke_
implementation_ppt

[37] ICAO. Airworthiness of aircraft. In:
International Standards and
Recommended Practices, Annex 8 to the
Convention on International Civil
Aviation. Montréal, Quebec, Canada:
ICAO; 2005

[38] ICAO. Safety Management Manual
(SMM). 2nd ed. Quebec, Canada:
Montréal; 2009

[39] FAA. AC 20-175. Controls for flight
deck systems. 2011

[40] Sun R, Zhao K, Zhang X. Research
on error-proofing design of Boeing and
airbus cockpit from pilots survey.
International Conference on Engineering
Psychology and Cognitive Ergonomics,
EPCE. 2015;**2015**:492-504

[41] Sianturi R, Suwondo E. The
development of a quantitative RCM/
MSG application using Microsoft Office.
International Journal of Aviation Science
and Engineering. 2020;**2**:57-66

[42] NPDSolutions. Mistake-Proofing by
Design. Florida, USA: NPD Solutions;
2019. Available from: https://www.npd-
solutions.com/mistake.html

[43] McBride D. Using mistake-proofing
in product design. Oklahoma, USA:
Noria Corporation; Available from:
https://www.reliableplant.com/Read/
13159/mistake-proofing-product-design

[44] Data4safety. Cologne, Germany:
Available from: https://www.easa.
europa.eu/en/domains/safety-
management/data4safety

[45] Member States. Montréal, Quebec,
Canada: ICAO; Available from: https://
www.icao.int/about-icao/Pages/
member-states.aspx

Chapter 2

A Methodological Framework for Certification by New Aircraft Integrators

Sophie Lemoussu, Rob Vingerhoeds, Pieter van Langen,
Frances Brazier and Jean-Charles Chaudemar

Abstract

Small and medium-sized enterprises (SMEs) are becoming more and more involved in designing and manufacturing new forms of aircraft, transitioning from the traditional role of parts supplier to system designer and integrator. This transition drastically changes the scope of their responsibility. As integrators, SMEs are responsible for the mandatory certification of the design capability, the manufacturing capability, and the aircraft components. This requires specialist knowledge of regulations, norms and standards, and dedicated procedures. As such, certification constitutes a major challenge for SMEs with limited experience and resources. To support European SMEs, this chapter presents an overview of a methodological framework for European Aviation Safety Agency (EASA) initial airworthiness certification. Developed in close cooperation with a French SME, the framework helps new aircraft integrators govern and execute certification processes. A case study with the French SME demonstrated its ability to perform certification by using the methodological framework.

Keywords: certification, process modelling, requirements modelling, small and medium-sized enterprises, systems engineering

1. Introduction

Aircraft certification is the process of assessing the quality and safety of an aircraft. The first part of an aircraft certification process, known as initial airworthiness, aims to ensure that any new aircraft is designed and manufactured with sufficient safety margins through acceptable, rigorous, and reproducible activities. The second part, known as continuing and continued airworthiness, aims to ensure that all operational conditions remain safe during the entire aircraft life cycle. The focus of this chapter is on the first part.

In Europe, requirements for initial airworthiness are listed in the European Aviation Safety Agency (EASA) regulatory framework Part 21 [1, 2]. Part 21 requires a manufacturer to apply for four certificates in sequence:

IntechOpen

- Design certificate, which takes into account a proposed design, a formal "type design", and the process followed. Usually, a Design Organisation Approval (DOA, Subpart J) is required, but in some situations, Alternative Procedures to DOA (APDOA, Subpart J) may suffice;

- Production certificate, which takes into account the production facility, given a DOA. This usually requires a Production Organisation Approval (POA, Subpart G);

- Type certificate (TC, Subpart B), which takes into account the approved "type design", given a Design certificate and a Production certificate;

- Certificate of Airworthiness (CoA, Subpart H), which ensures that the assembly of the aircraft conforms to the original approved "type design" with respect to the expected operational conditions.

These certificates are mandatory for each "type design" of an aircraft to enter the market. Each certificate has its own set of requirements and other rules to fulfil, such as Acceptable Means of Compliance (AMCs) and Guide Materials (GMs), adapted according to the industrial context. For a well-established original equipment manufacturer (OEM) with a strong foundation and experience in certification activities, the list of certification requirements may be different that for a new entrant such as a small or medium-sized enterprise (SME).

SMEs in the European aeronautical industry develop innovative aircraft, such as airships and electric Vertical Take-Off and Landing (eVTOL) aircraft. In practice, understanding the certification process itself is difficult enough for SMEs, given their limited experience and resources. Once understood, certification requires more detailed specifications of work processes and performance indicators to monitor expected quality, such as design assurance processes and quality assurance processes [3]. Moreover, most requirements of Part 21, AMCs, and GMs are related to the internal organisation of an enterprise and impose a process orientation towards quality and design assurance.

This chapter introduces a model-based approach to support European SMEs in the certification of new aircraft. A methodological framework helps new aircraft integrators govern and execute initial airworthiness certification processes through models. It relieves them of the burden to study and grasp the essentials of a huge number of norms and regulations. The origins and details of this research have been presented in Refs. [4, 5].

This chapter is structured as follows. Section 2 provides background information on the development of a methodological framework and identifies the need to combine business process modelling and requirements modelling for certification purposes. Section 3 presents a methodological framework for EASA initial airworthiness certification. Section 4 presents a case study with a French SME from the aeronautical industry, where the methodological framework has been developed. Section 5 concludes the chapter, proposing to extend the approach to other certification processes and industries.

2. Background

Systems engineering approaches provide a wide collection of methods and processes in order to tackle the complexity of various types of systems. Rigorous and interdisciplinary methodologies are needed to manage their design, development, validation, and verification during their entire life cycle. This chapter takes a model-based systems engineering (MBSE) approach.

2.1 Model-based systems engineering

Model-based systems engineering (MBSE) supports system life cycle phases and activities through often graphics-based modelling and analysis tools. This sub-section briefly discusses a framework for MBSE.

Architecture Analysis and Design Integrated Approach (ARCADIA) is a structured MBSE method developed by Thales Group for system, hardware, and software architectural design [6]. Supported by the open-source Capella tool, ARCADIA places the system architecture and the interactions between actors and entities at the centre of system design [7]. Another approach is the Object Process Methodology (OPM) [8, 9], which has increasing popularity. OPM builds on objects, processes, and their links, representing a system simultaneously in formal graphics and natural language. Both OPM and ARCADIA try to achieve better-structured systems engineering approaches while remaining fully compliant with systems engineering standards.

2.2 Systems modelling

MBSE is commonly associated with SysML, the systems modelling language standardised by Object Management Group (OMG)[1] with the support of International Council on Systems Engineering (INCOSE). SysML is a graphic and semi-formal modelling language that has been defined independently of any tool or associated method and applied to a broad variety of systems. SysML tools include model editors, model simulators, model checkers, code generators, and test sequence generators. At the time of writing this chapter, version 2 of SysML is in preparation, but not yet officially standardised. SysML is used in this chapter specifically to integrate multiple views in system descriptions.

2.3 Requirements modelling

In systems engineering design, requirements are often inconsistent, incomplete, and ambiguous. This is also a challenge for the EASA initial airworthiness certification process, which involves many requirements of different types. To structure these requirements, this chapter deploys an Function-Behaviour-Structure (FBSE) requirements classification framework based on [10, 11] that distinguishes three types of requirements (**Table 1**).

3. A methodological framework for certification

This section presents a methodological framework for EASA initial airworthiness certification. The framework provides new aircraft integrators with support to govern and execute certification processes. **Figure 1** describes the main components of the framework: certification governance, a meta-model for certification processes and a reference model for certification requirements.

---

[1] See Available from: https://www.omgsysml.org/.

Requirements class	Meaning	Questions answered
Function	The purpose of a system	Why? For whom? Where?
Behaviour	The way a system acts	How? When?
Structure	The components of a system and their relationships	What?

Table 1.
FBS requirements classification framework.

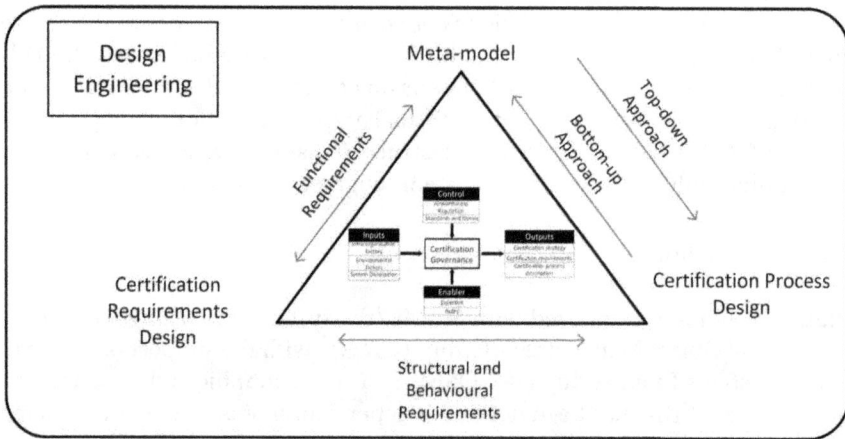

Figure 1.
Methodological framework for certification.

According to this framework, certification follows three main steps as described in **Figure 2**: designing a certification strategy, designing certification requirements, and designing certification processes.

3.1 Governance of certification

Governance determines both organisational and business strategies [10]. Regarding certification, it aims to help enterprises anticipate the strategic decisions needed to acquire the necessary certificates (**Figure 3**). Governance of certification is defined as a decision-making process starting from context analysis, constituting a set of knowledge needed to generate a dedicated certification strategy, certification requirements, and a certification process description. The necessary knowledge to govern the initial airworthiness certification of a new aircraft is accounted for as "Inputs":

- *Intra-organisation factors*: Organisational characteristics such as the size of the enterprise (i.e., the total number of employees), the types of activity of the enterprise, and its market strategy;

- *Environmental factors*: Characteristics of the system to be developed, such as the type of aircraft, its weight, and its operational conditions;

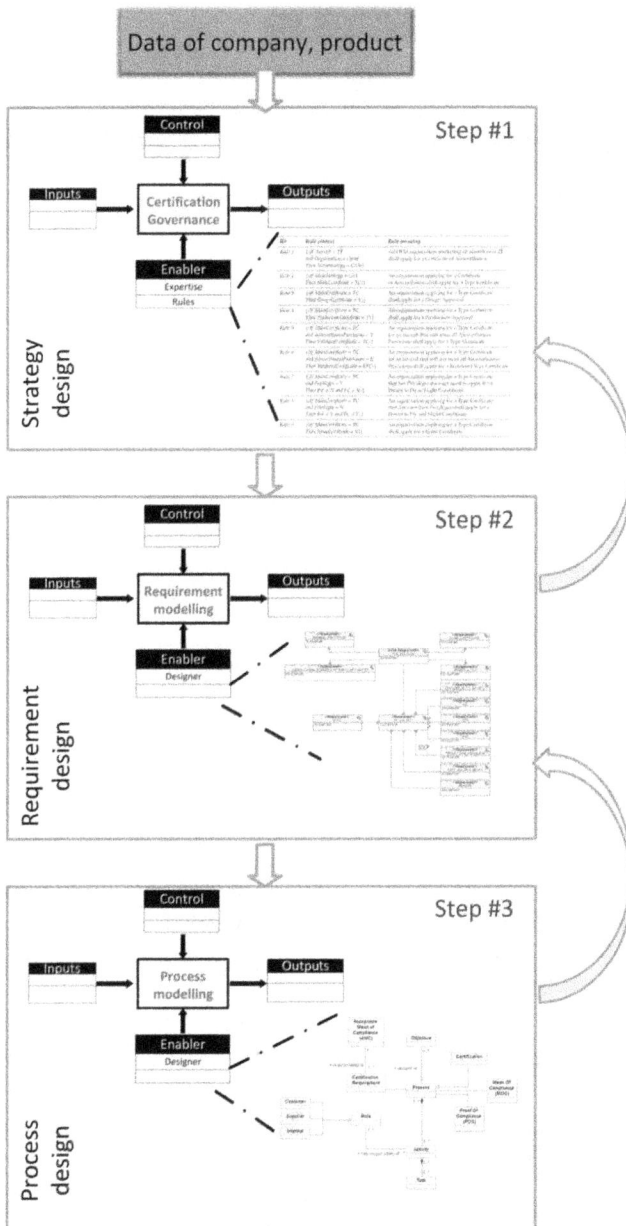

Figure 2.
Overview of the certification method.

- *System description*: Properties of the design and production capability, such as demonstrating airworthiness provision and having privileges.

The main goal of governance in certification is to generate a certification strategy: a list of sub-part certificates to apply for in the context of the enterprise. The choice of strategy will result in a list of case-specific certification requirements and a case-specific process description. This is simple enough to be understood well by an organisation without any

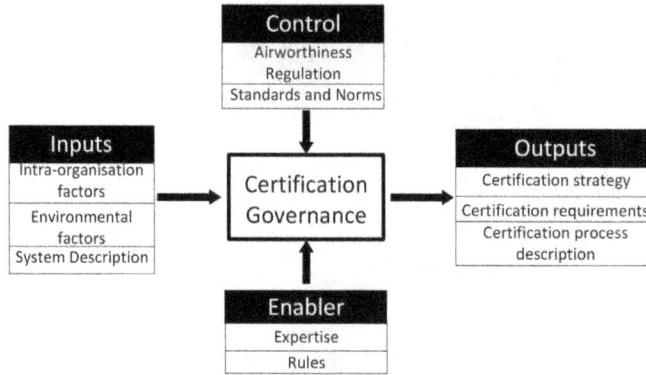

Figure 3.
Governance of certification.

ID	Rule content	Rule meaning
Rule 1	{|*If Aircraft* > 2*T* *and Organisation* = *OEM* *Then MainStrategy* = *CoA*|}	An OEM organisation marketing an aircraft over 2 T shall apply for a Certificate of Airworthiness
Rule 2	{|*If MainStrategy* = *CoA* *Then MainCertificate* = *TC*|}	An organisation applying for a Certificate of Airworthiness shall apply for a Type Certificate
Rule 3	{|*If MainCertificate* = *TC* *Then DesignCertificate* = *Y*|}	An organisation applying for a Type Certificate shall apply for a Design Approval
Rule 4	{|*If MainCertificate* = *TC* *Then ProductionCertificate* = *Y*|}	An organisation applying for a Type Certificate shall apply for a Production Approval
Rule 5	{|*If MainCertificate* = *TC* *and AirworthinessProvisions* = *Y* *Then ValidatedCertificate* = *TC*|}	An organisation applying for a Type Certificate for an aircraft that will meet all Airworthiness Provisions shall apply for a Type Certificate
Rule 6	{|*If MainCertificate* = *TC* *and AirworthinessProvisions* = *N* *Then ValidatedCertificate* = *RTC*|}	An organisation applying for a Type Certificate for an aircraft that will not meet all Airworthiness Provisions shall apply for a Restricted Type Certificate
Rule 7	{|*If MainCertificate* = *TC* *and Privileges* = *Y* *Then PtF* = *N and FC* = *N*|}	An organisation applying for a Type Certificate that has Privileges does not need to apply for a Permit to Fly or Flight Conditions
Rule 8	{|*If MainCertificate* = *TC* *and Privileges* = *N* *Then PtF* = *Y and FC* = *Y*|}	An organisation applying for a Type Certificate that does not have Privileges shall apply for a Permit to Fly and Flight Conditions
Rule 9	{|*If MainCertificate* = *TC* *Then NoiseCertificate* = *Y*|}	An organisation applying for a Type Certificate shall apply for a Noise Certificate

Table 2.
Decision rules for governance of certification.

specific training. Furthermore, it promotes a better understanding of the relations between initial airworthiness and the certification requirements to be satisfied.

Generating a certification strategy can be supported by an expert system representing all sub-part certificates as decision rules [12]: mappings from a set of conditions to outcomes. Then, governance of certification can be viewed as an inference engine using a knowledge base and a set of facts to generate a certification strategy. Part of the decision rules that the knowledge base contains are listed in **Table 2**.

3.2 Certification requirements design

For the creation of certification requirements, the methodological framework provides a certification requirements reference model. A requirements diagram with the main sub-parts of Part 21 and their relationships is depicted in **Figure 4**.

- Sub-part A: General provisions

- Sub-part B: Type Certificates (TCs) and Restricted TCs

- Sub-part F: Production without Production Organisation Approval (POA)

- Sub-part G: POA for products, parts, and appliances

- Sub-part J: Design Organisation Approval (DOA)

- Sub-part K: Parts and appliances

- Sub-part M: Repairs

- Sub-part O: European Technical Standard Order (ETSO)

- Sub-part Q: Identification of products, parts, and appliances

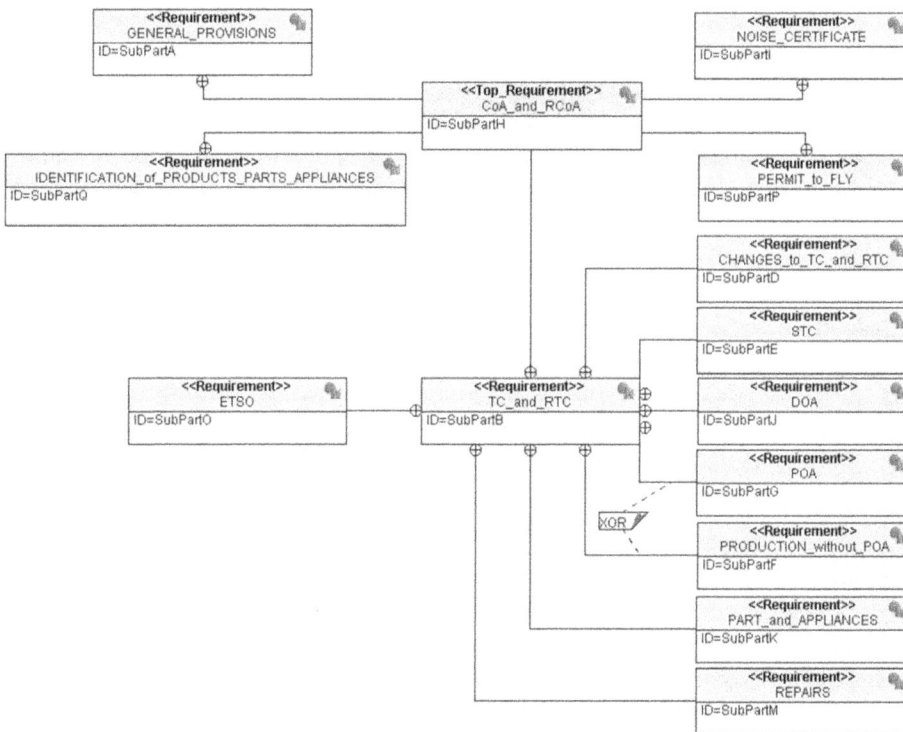

Figure 4.
Top-level certification requirements diagram for an aircraft manufacturer.

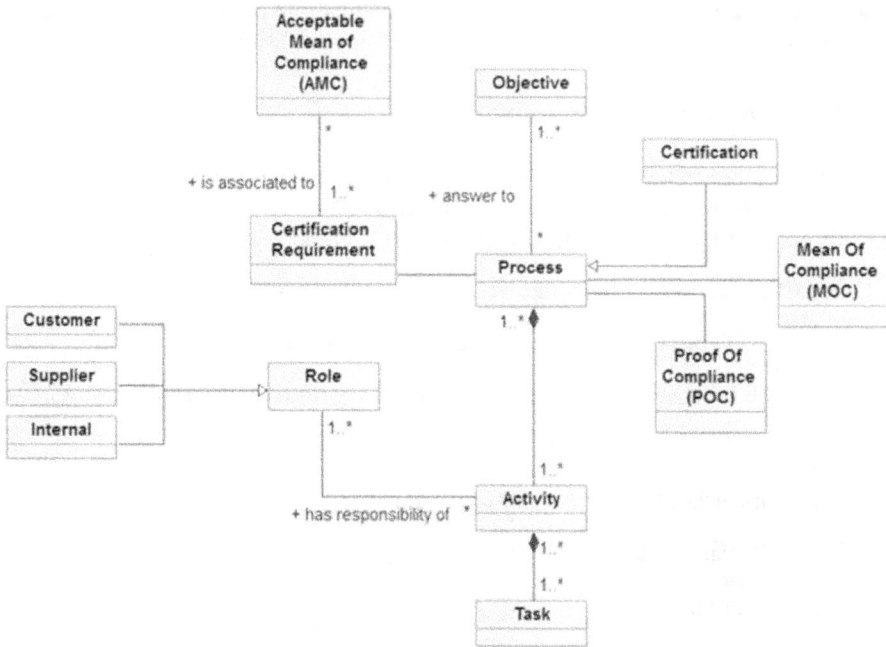

Figure 5.
Certification process meta-model.

3.3 Certification process design

For certification, the organisation of people, processes, and products is of importance. **Figure 5** shows a meta-model of certification expressed in SysML. The meta-model distinguishes tasks, activities, and processes, together with roles (responsible for activities), datasets, and contexts.

4. Case study

A case study has been conducted at a French SME that has contributed significantly to the research. This company is developing a new form of air transportation with several electric engines. A few years ago, the enterprise employed fewer than 10 people, whereas it now around e80 employees. With a new, non-conventional supply chain and very little aeronautical experience, the enterprise is facing many challenges, including those related to the organisation itself, requiring almost continual change management. The certification process is one of the major challenges, and the company explicitly requested support for this activity.

Following the first step of the methodological framework, a certification strategy has been generated. **Table 3** depicts this strategy as a list of sub-part certificates to apply for, with, for each, the number of corresponding requirements and also the number of associated AMCs and GMs, which must be considered equally important requirements to be taken into account, even if they are not strictly regulatory requirements.

Subpart	Title	Reqs	AMCs	GMs
A	General provisions	5	5	4
B	Type Certificate (TC) and restricted TC	24	4	17
G	Production Organisation Approval (POA)	17	14	27
H	Certificate of Airworthiness (CoA) and Restricted CoA	10		
I	Noise Certificates	7		
J	Design Organisation (DOA)	16	13	15
K	Parts and appliances	4	1	1
M	Repairs	14	4	10
P	Permit to fly	16	1	15
Q	Identification of products, parts, and appliances	5		1
Total		118	42	90

Table 3.
List of sub-part certificates to apply for and the number of applicable requirements in the case study.

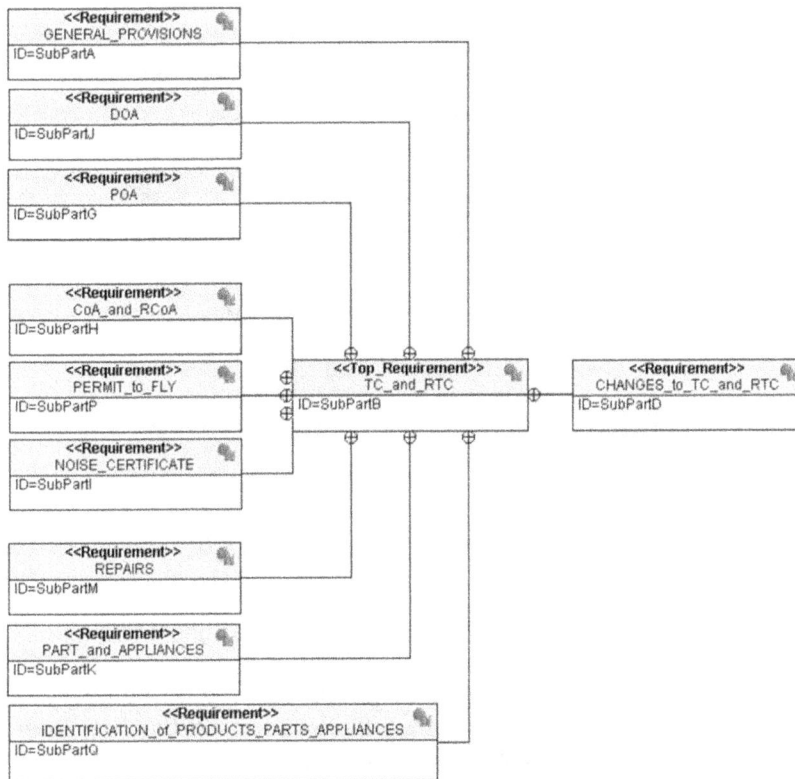

Figure 6.
Top-level certification requirements in the case study.

Process	Activity	Task
DOA management	Apply DOA	Produce design handbook
	Maintain DOA	Update design handbook
	Monitor and control DOA	
POA management	Apply POA	Produce production handbook
	Maintain POA	Update production handbook
	Monitor and control POA	
TC management	Apply TC	Prepare certification programme
	Maintain TC	Prepare compliance checklist
	Monitor and control TC	Update compliance checklist
CoA management	Apply CoA	Provide certification basis
	Maintain CoA	
	Monitor and control CoA	
Design change management	Classify design change	
	Manage minor change	Apply for minor change approval
	Manage major change	Apply for major change approval
Design assurance	Plan design activities	Plan systematic actions
	Maintain design activities	Update planned actions
	Manage corrective actions	
Independent monitoring	Plan monitoring activities	Perform audits
	Perform monitoring activities	Control and monitor the design
	Manage findings	
Compliance demonstration	Manage compliance documents	Produce compliance documents
	Analyse compliance documents	Update compliance documents
	Declare compliance	
Liaison with authorities	Manage findings	Manage handbook applications
	Ensure reporting	Manage approval changes
	Manage data	Manage deliveries

Table 4.
Process structure in the case study (non-exhaustive).

Following the second step, certification requirements have been generated, using the reference model from the methodological framework (**Figure 6**).

Following the third step, a model of the certification process to execute the certification strategy has been built, using the certification process meta-model from the methodological framework. **Table 4** shows activities and tasks that have been generated for the identified requirements in the second step.

The qualitative results have been encouraging, the SME has shown to have a better understanding of the objectives of certification, and is less burdened by the large number of norms and standards.

5. Conclusions

Certification in the aerospace industry should support innovative design by SMEs that are new to the role of integrator. This chapter introduced a methodological framework enabling an SME to extract applicable certification requirements and generate a customised certification process. Contrary to current approaches, the process is the central element of our approach. The methodological framework consists of a unique and common model-based environment in which both requirements and processes are specified explicitly, providing the details needed for certification.

One direction for future research is to further explore cases and other types of certification in the aeronautical industry. This will lead to adjustments and extensions of the methodological framework. Furthermore, the approach may be adapted, with the necessary efforts, to any context where regulation imposes numerous requirements and impacts the development processes. Therefore, another direction for future research is to explore certification in other industries.

Conflict of interest

The authors declare no conflict of interest.

Thanks

This work was supported by the Defense Innovation Agency (AID) of the French Ministry of Defense (research project called CONCORDE No 2019 650090004707501).

Nomenclature

In **Table 2**.

FC	flight conditions
N	negative answer NO
PtF	permit to fly
Y	positive answer YES

Abbreviations

AMC	acceptable means of compliance
APDOA	alternative procedures to DOA
ARCADIA	architecture analysis and design integrated approach
CoA	certificate of airworthiness
DOA	design organisation approval
EASA	European aviation safety agency
ETSO	European technical standard order
eVTOL	electric vertical take-off and landing
GM	guidance material
INCOSE	international council on systems engineering

MBSE	model-based systems engineering
OEM	original equipment manufacturer
OMG	object management group
OPM	object process methodology
POA	production organisation approval
SME	small and medium-sized enterprise
SysML	systems modelling language
TC	type certificate

Author details

Sophie Lemoussu[1], Rob Vingerhoeds[1], Pieter van Langen[2], Frances Brazier[2] and Jean-Charles Chaudemar[1*]

1 ISAE-SUPAERO, Toulouse, France

2 Delft University of Technology, Delft, The Netherlands

*Address all correspondence to: jean-charles.chaudemar@isae-supaero.fr

IntechOpen

© 2025 The Author(s). Licensee IntechOpen. This chapter is distributed under the terms of the Creative Commons Attribution License (http://creativecommons.org/licenses/by/4.0), which permits unrestricted use, distribution, and reproduction in any medium, provided the original work is properly cited. (cc) BY

References

[1] EASA Certificates and Approvals; 2018. Available from: https://www.easa. europa.eu/document-library/ application-forms/certificates-and-approvals [Accessed: November 14, 2018]

[2] EASA. Easy access rules: Airworthiness and environmental certification (regulation (EU) No 748/ 2012). In: Aircraft System Safety. European Union; 2018. pp. 193-324

[3] Kritzinger D. Aircraft system safety: Assessments for initial airworthiness certification. Duxford, United Kingdom: Woodhead Publishing; 2016. pp. 193-324

[4] Lemoussu S, Chaudemar JC, Vingerhoeds RA. Systems engineering and Project Management process Modeling in the aeronautics context: Case study of SMEs. International Journal of Mechanical and Mechatronics Engineering. 2018;**12**:88-96

[5] Lemoussu S. A Model-Based Framework for Innovative Small and Medium-sized Enterprises (SMEs) in Aeronautics; 2020. Thèse de doctorat dirigée par Vingerhoeds, Rob A. et Chaudemar, Jean-Charles Génie Industriel - Automatique et informatique Toulouse, ISAE. 2020. Available from: http://www.theses.fr/2020ESAE0014

[6] Normand V, Exertier D. Model-driven systems engineering: SysML and the MDSysE approach at Thales. In: Model Driven Engineering for Distributed Real-Time Embedded Systems. London, United Kingdom: ISTE Ltd; 2005

[7] Roques P. MBSE with the ARCADIA method and the Capella tool. In: 8th European Congress on Embedded Real Time Software and Systems (ERTS 2016). 2016

[8] Dori D. Model-Based Systems Engineering with OPM and SysML. New York, United States of America: Springer; 2016. pp. 407-422

[9] ISO ISO/IEC 19514:2017 - Information Technology — Object Management Group Systems Modeling Language (OMG SysML); 2017

[10] Gero JS, Kannengiesser U. A function–behavior–structure ontology of processes. In: Design Computing and Cognition '06. Dordrecht, Netherlands: Springer; 2006. pp. 407-422

[11] Brazier FM, Van Langen PHG, Lukosch S, Vingerhoeds RA. Complex systems: Design, engineering, governance. In: No. 104 in Projects and People: Mastering Success. Nijkerk, Netherlands: NAP; 2018

[12] Liao SH. Expert system methodologies and applications—A decade review from 1995 to 2004. Expert Systems with Applications. 2005;**28**: 93-103

Chapter 3

The Relationship between Turbulence-Related Aircraft Accidents and Geomagnetic Storms in Civil Aviation

Ümit Deniz Göker

Abstract

Turbulence is an aviation phenomenon that generally does not result in fatalities and is caused by many different reasons. However, our research has shown that turbulence resulting in death did occur. In this study, fatal aircraft accidents in civil aviation caused by turbulence between the years 1928 and 2024 are investigated and found that geomagnetic storms (GSs) also happened during these accidents. Of all the aircraft accidents resulting in death during this period, only 35 accidents were caused by turbulence, and the percentage of their occurrence at geographic latitudes of 40% ($\geq$40°), 37% (30° $\leq \varphi$ < 40°), and 23% (<30°), respectively, but these accidents were preceded by GSs is estimated. The distribution of the intensity of these GSs was moderate (45%), weak (42%), and strong/very strong (12%). It is also seen that the frequency of occurrence of these accidents increased during the decreasing phases or the ending of the increasing phases of solar activity cycles. This chapter mainly focuses on the relationship between turbulence-related aircraft accidents and GSs but also discusses aircraft safety and management practices to minimize these accidents.

Keywords: aircraft accidents, aircraft safety, civil aviation, geomagnetic storms, safety management, turbulence

1. Introduction

Aviation safety is the study and practice of managing risks in aviation, which includes preventing aviation accidents and incidents through research, education in air travel personnel, passengers, etc., as well as the design of aircraft and aviation infrastructure. In this work, we focus on aircraft accidents resulting in death between the years 1928 and 2024 depending on the turbulence in civil aviation and possible aircraft safety methods that can be applied by air travel personnel from ground to flight (e.g.. ground personnel, pilot, and cabin crew) directly and safety management systems by organizations.

Turbulence is a weather-related unpredictable aviation phenomenon that is caused by many different reasons and generally does not result in fatalities, and it changes

IntechOpen

the altitude, attitude, and/or airspeed of aircraft depending on intensity. Turbulence is responsible for approximately 30% of all weather-related aircraft accidents, and it mainly occurs when an aircraft hits a strong wind current. Erratic motion of air is another reason for turbulence, and it works out from obstructions to the airflow (mechanical) or vertical currents (convective). Turbulence is classified as *light* (unpredictable rhythmic slight changes in altitude and/or attitude (e.g., pitch, roll, or yaw)), *moderate* (variations in airspeed, rapid bumps or jolts but positive control of aircraft during the changes in altitude and/or attitude), *severe* (sudden and effective changes in altitude and/or attitude cause strong variations in airspeed and aircraft may be instantly out of control), and *extreme* (the aircraft is extremely staggered without control and it can cause structural damage) [1, 2].

There are many environmental components that cause turbulence: (1) *thermal (convective) turbulence*: turbulent vertical motions of air which are caused by unstable heating of the Earth's surface, especially warm summer afternoons, lead to ascending and descending currents of air at lower altitudes and causes strong up- and down-drafts; (2) *wake turbulence*: it is caused by going after a larger aircraft on take-off or landing, and wingtip vortices that are spiraling turbulent air are created by the meeting of two different pressures on the wing's profile as the lower and higher pressures on the upper and lower surfaces of the wing meet each other; (3) *windshear*: aircraft lift performance is affected by sudden movement in wind speed or direction in a short distance and it is mainly effective at lower altitudes; (4) *frontal turbulence*: weather fronts create the frontal turbulence with the collision of cold and warm air masses and the colder air rises upwards to replace the warmer air. It is mostly associated with cold fronts, while warm fronts are important to a lesser degree. This turbulence relates to the thunderstorms, as well; (5) *thunderstorm turbulence*: aircraft can be destroyed due to strong up- and downdrafts within the storm cells which can be extreme inside or near thunderstorms; (6) *mechanical turbulence*: uneven terrain and man-made (e.g., huge hangars and a row of trees) barriers to the wind flow have affected the aircraft depending on the strength of the wind, and turbulence is higher when the wind flow directly perpendicular to the runway; (7) *mountain wave turbulence*: it has a similar effect as in mechanical turbulence but in a larger scales, and the wind blows directly perpendicular to a mountain range. The air blows across the mountain from downwards to upwards and encounters the wind on its way up the mountain. As a result, it creates an irregular airflow where the wind is blowing, and the airflow on the windward side of the mountain is further disrupted; and (8) *clear air turbulence* (*CAT*): CAT occurs as a sudden severe turbulence when the jet stream interacts with the surrounding calm air in cloudless regions in higher altitudes, generally above 15,000 feet. It is dangerous due to the lack of visual cues of its presence, and it is not possible to detect CAT with the weather radar onboard advanced aircraft [1]. All these turbulences are shown in **Figure 1**.

There are many aviation organizations that consider aircraft safety and safety management programs such as the International Civil Aviation Organization (ICAO), the U.S. Federal Aviation Administration (FAA), the European Aviation Safety Agency (EASA), Transport Canada Civil Aviation (TCCA), Civil Aviation Authority of New Zealand (CAANZ), Civil Aviation Safety Agency (CASA) of Australia, Agência Nacional de Aviação Civil (ANAC), Brazil, Federal Office of Civil Aviation (FOCA), Switzerland, Direction Générale de l'Aviation Civile (DGAC), France, United Kingdom Civil Aviation Authority (UK CAA), and Eurocontrol, Flight Safety Foundation (FSF). Safety management programs also involve the determination and evaluation of safety issues, identification and programming of safety actions,

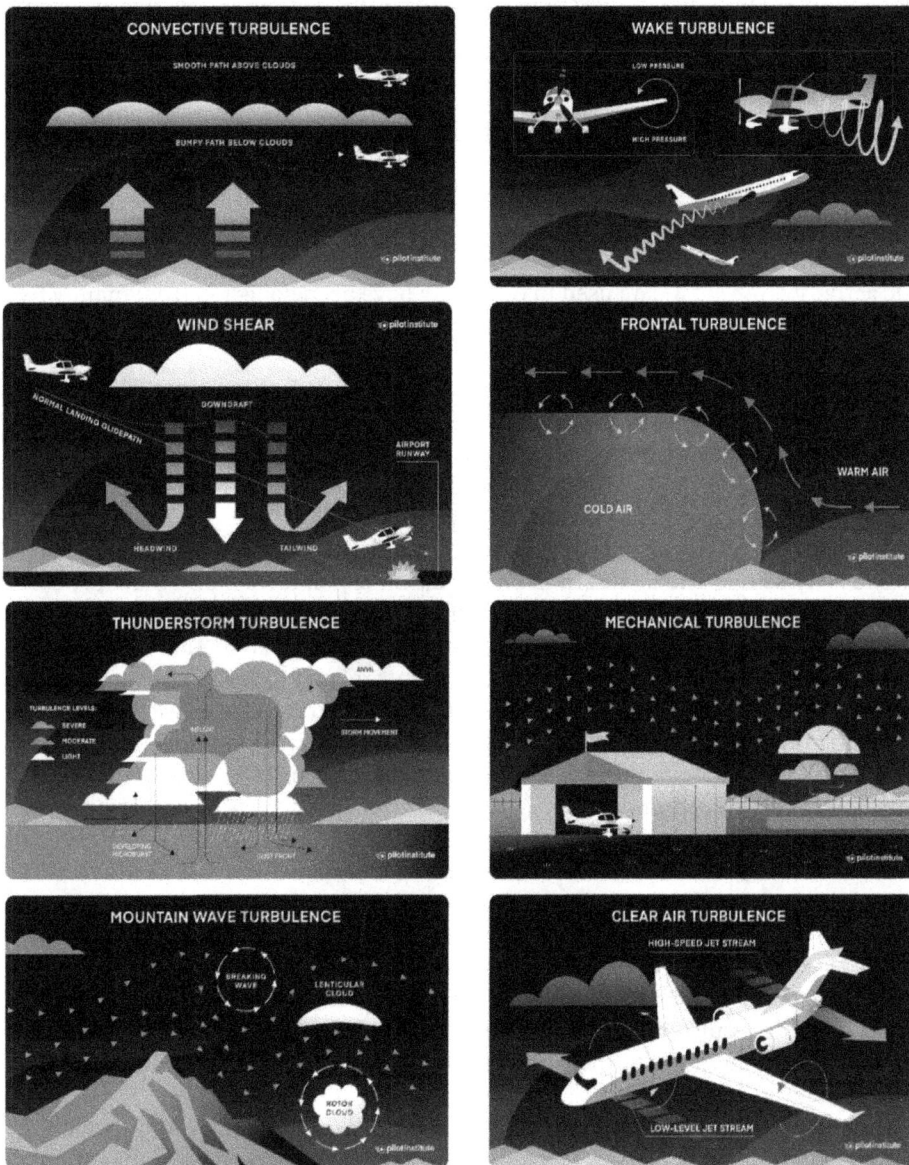

Figure 1.
Different types of turbulence [1].

performing and follow-up, and safety performance measurement. The State Safety Programme (SSP) is an integrated set of regulations and activities aimed at improving safety (ICAO Annex 19) [3].

On a large scale, safety policy includes the fundamental approach of an organization to managing safety that is to be adopted within an organization and further defines the organization management's commitment to safety and overall safety vision (FAA AC 150/5200-37) while the risk management process describes a system that assesses hazards, analyses those hazards to evaluate the risk, and establishes controls to manage those risks on a small scale [3]. Based on this description, the risk factor we mentioned in this study is *only* for turbulence, and the potential safety procedures to minimize this risk factor are also given in detail.

2. Turbulence-related aircraft accidents

We analyzed air crash investigations (ACI) from the decreasing phase of the solar cycle (SC) 16 to the increasing phase of SC 25 and investigated the effect of geomagnetic storms (GSs) depending on seasonal changes, solar activity (SA), and geomagnetic activity (GA). ACI data used in this study have been collected from mainly two sources, that is, https://www.planecrashinfo.com/ and https://www.ntsb.gov/safety/data/Pages/Data_Stats.aspx [4, 5].

We included all civil and commercial ACI of scheduled and non-scheduled passenger airlines worldwide in the database. We dealt with all cargo and ferry accidents that resulted in a fatality, but we did not consider military transport and test flight fatal accidents. These data are determined and drawn by using "MATLAB R2018a Programming Language" and shown in **Figures 2** and **3**.

We extracted fatal and turbulence-related accidents from the 1959 ACI we had previously classified and analyzed in civil aviation [6], and we found that only 35 accidents were caused by turbulence. Information on all the aforementioned aircraft accidents between January 21, 1928, and May 21, 2024, is given in **Table 1**. In addition to this, GSs and their corresponding accidents related to turbulence are matched both for increasing and decreasing phases of solar activity cycles (SACs) 16–24, and a comparison of the number of accidents with the GA intensities is given in **Figures 2** and **3**. These figures show that the accident rates of 1959 air crash events that might be related to GSs correspond to 68, 22, 8, and 2% for mechanical and pilotage reasons, unknown reasons, bad weather conditions, and turbulence, respectively. We obtained these data for latitudes between $\pm 10° \leq \varphi \leq \pm 90°$ in the northern and southern hemispheres because the most severe storms may have an impact as low as $\pm 10°$ depending on their higher intensity values [6]. It is important to note here that the GS activity is more effective at higher geographic latitudes than at lower latitudes. GSs have a devastating impact on the magnetosphere of Earth in precise periods, and they cause damaging results on mechanical/technical and electronic accessories of aircraft and mental and physical health problems for pilot and cabin crew, especially at high latitudes flights [7].

Solar-induced effects that cause a GS are *coronal mass ejections* (*CMEs*) which generate high-energetic particles including electrons and coronal/SW ions (mainly protons), *coronal holes* (*CHs*), the classes of *solar flares* as C-, M-, or X-, the variable duration of the SC and increasing and decreasing phases of the SC, and the *solar winds* (*SWs*) propagating through shock waves that transmit these effects toward the Earth [6]. The effectiveness of SWs continues even after 24–36 hours from the moment they interact with the Earth's magnetosphere.

The irregular air movement, which is called turbulence, occurs for many reasons, and one of these reasons is the change in atmospheric pressure because of differences in energy flux from the Sun; especially during GSs, this pressure change is very high, and the turbulence effect also increases [8]. After the eruption, highly energetic particles from CMEs and/or CHs reach the Earth between 2 and 6 days, depending on the intensity of the GS; however, the most severe storms reach the Earth in 1 day. The results of extreme space weather events and low frequency/high effect superstorms

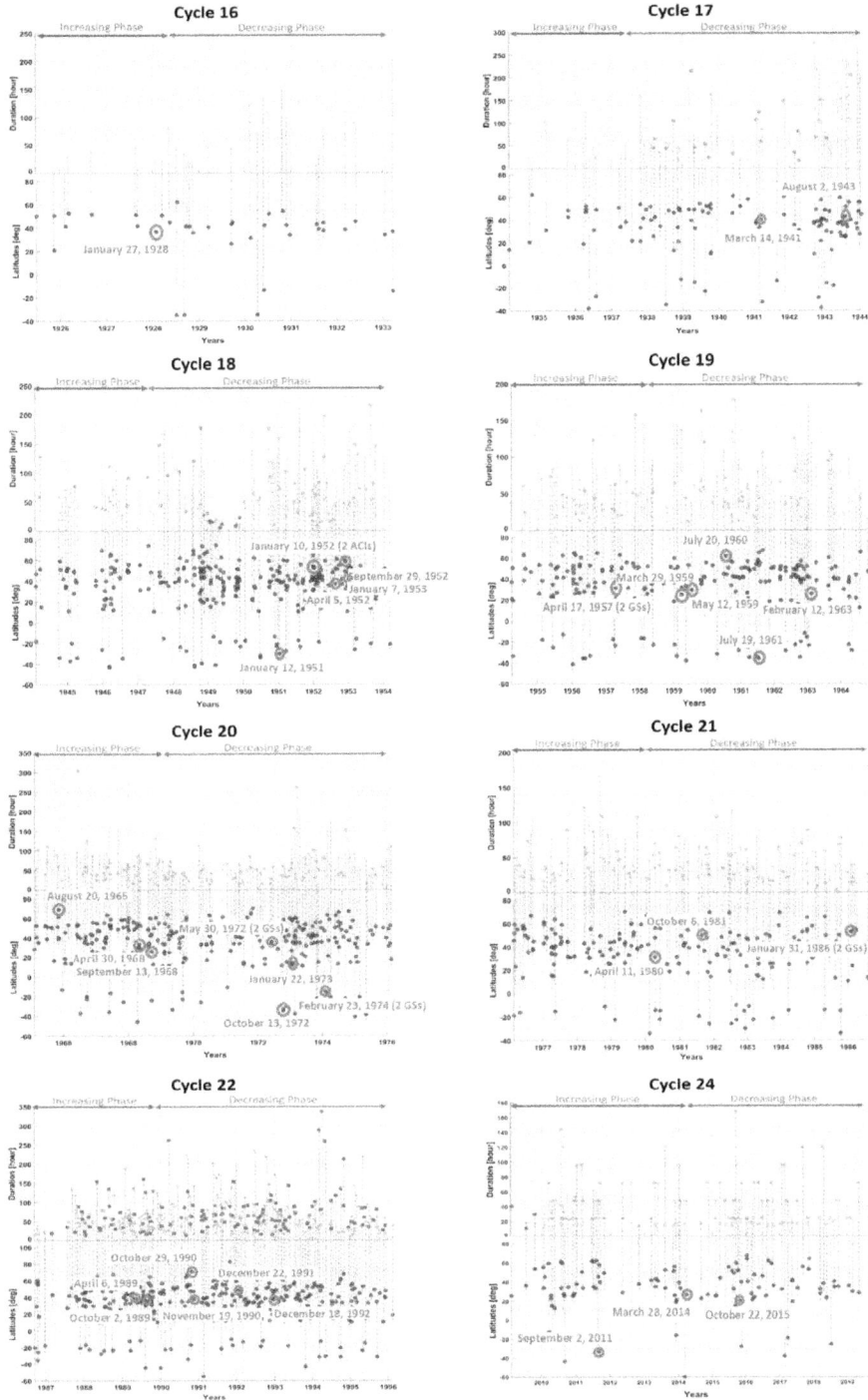

Figure 2.
GSs and their corresponding accidents related to turbulence are matched both for increasing and decreasing phases of SACs 16–24. These plots are taken from the paper of [6], and turbulence-related accidents are signed with circles on these plots. We did not match any ACI only for the SC 23, even with its longer period of activity cycle. We also did not add cycle 25 because there was only one accident related to turbulence on May 21, 2024, and the SC was not completed.

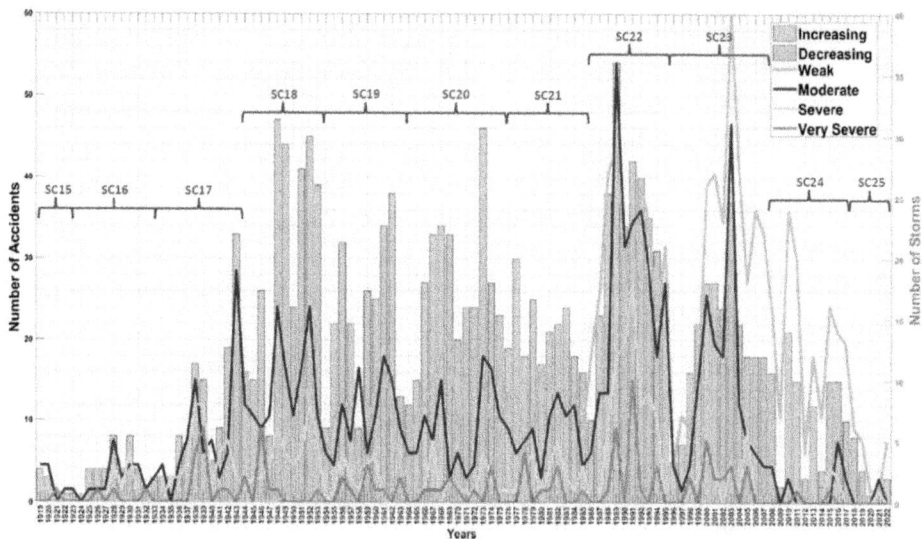

Figure 3.
Comparison of the number of accidents that occurred during the increasing and decreasing phases of the SACs from Cycle 15 to Cycle 25 with the GA intensities [6].

Accident data	Latitude/longitude	Geomagnetic data	Intensity
27/01/1928	36°50′44″, −02°26′45″	26/01/1928	Moderate
14/03/1941	40°13′26″, 69°15′15″	13/03/1941	Strong
02/08/1943	43°12′25″, −93°05′05″	02/04/1943	Moderate
12/01/1951	−30°09′08″, 30°03′58″	10/01/1951	Weak
10/01/1952	52°23′14″, −03°52′52″	09/01/1952	Moderate
10/01/1952	53°11′00″, −08°01′53″	09/01/1952	Moderate
05/04/1952	40°44′31″, −74°09′50″	30/03/1952	Strong
29/09/1952	55°02′11″, 88°28′28″	28/09/1952	Moderate
07/01/1953	42°02′17″, −111°23′23″	05/01/1953	Moderate
17/04/1957	31°08′40″, 36°40′40″	15/04/1957	Weak
(2 GSs)		17/04/1957	Strong
29/03/1959	24°49′57″, 92°46′46″	26/03/1959	Very strong
12/05/1959	33°04′07″, −96°47′47″	11/05/1959	Strong
20/07/1960	61°39′23″, 50°50′50″	19/07/1960	Weak
19/07/1961	−36°14′35″, −59°22′22″	17/07/1961	Moderate
12/02/1963	25°48′31″, −80°35′35″	09/02/1963	Moderate
20/08/1965	50°44′06″, 05°12′12″	18/08/1965	Moderate
30/04/1968	33°40′29″, 73°08′34″	25/04/1968	Moderate
13/09/1968	26°01′51″, −80°10′10″	12/09/0968	Moderate
30/05/1972	32°44′09″, −97°21′21″	28/05/1972	Weak

Accident data	Latitude/longitude	Geomagnetic data	Intensity
(2 GSs)		30/05/1972	Weak
13/10/1972	−34°35′23″, −70°59′14″	12/10/1972	Weak
22/01/1973	11°58′09″, 08°31′44″	19/01/1973	Weak
23/02/1974	−16°40′00″, −65°11′11″	20/02/1974	Weak
(2 GSs)		22/02/1974	Moderate
11/04/1980	31°20′48″, −98°56′56″	11/04/1980	Weak
06/10/1981	51°41′23″, 04°36′39″	02/10/1981	Moderate
31/01/1986	52°49′46″, −01°19′19″	27/01/1986	Moderate
(2 GSs)		29/01/1986	Weak
06/04/1989	39°50′09″, −88°52′52″	03/04/1989	Moderate
02/10/1989	34°51′14″, −111°47′47″	30/09/1989	Weak
29/10/1990	71°00′34″, 25°59′59″	26/10/1990	Weak
19/11/1990	37°58′59″, 58°16′06″	16/11/1990	Moderate
22/12/1991	49°26′01″, 08°41′41″	20/12/1991	Weak
18/12/1992	45°48′13″, −108°32′32″	17/12/1992	Moderate
02/09/2011	−33°3333″, −70°39′16″	28/08/2011	Weak
28/03/2014	26°29′16″, 77°00′00″	22/03/2014	Weak
22/10/2015	20°27′32″, −100°37′37″	21/10/2015	Weak
21/05/2024	28°31′50″, 97°37′55″	17/05/2024	Moderate
(2 GSs)		19/05/2024	Weak

A detailed explanation of data collection for geomagnetic data and intensity that are measured by ground-based observatories and satellites is given in our paper [6].

Table 1.
Information on all the fatal and turbulence-related ACI and corresponding GSs are given [4–6].

spread out from the solar corona are enormous on an extensive scale, from power supplies, navigation, satellites, the aviation industry, monitoring, and radio communications to social and economic effects on Earth.

We found in our analysis that the number of turbulence-related aircraft accidents that resulted in death from higher to lower occurred in January, April, and October and in the equinoxes in March and September; and in February, July, August, and December in the second order, and November is the month that only one accident was analyzed. In addition to this, the highest number of turbulence-related aircraft accidents seen is SC 20. The other cycles that follow from higher to lower are SC 19, SC 22, SC 18, SC 21, and SC 24, respectively. The only SC without any such kind of accident is SC 23. The most important factor in the turbulence-related and fatal aircraft accidents mentioned in our study is that the cause of the turbulence has not yet been clarified; that is, the National Transportation Safety Board (NTSB) still has not determined what caused the accidents, even though much time has passed. Therefore, we thought it would be most appropriate to attribute the cause of these accidents to GSs because we do not have a clear accident report yet.

3. The variation of GSs along with the turbulence-related aircraft accidents

GSs can be created by CMEs and/or CHs in different periods, and the SW plasma carries the magnetic fields (MFs) generated during GSs. The propagation of MFs from the Sun to the Earth's heliosphere is multi-variate in the solar maximum than at the solar minimum. Thus, the high-energetic particles from the Sun expand by the irregularities of the MF in the interplanetary medium reaching the top of the atmosphere is higher at a solar minimum than at a solar maximum. Contrary to this, the flux of galactic high-energetic particles enters the Earth's atmosphere at much larger scales in the minimum phase of SA because CHs are much smaller, short-lived, and more numerous at lower latitudes near the solar maximum; however, CHs are larger, longer-lived and fewer at high latitudes but higher number in equatorial and mid-latitudes near the solar minimum. This is because the angular sizes of CHs increase from the poles to equatorial and mid-latitude regions depending on the variation of open MFs from the solar maximum to the solar minimum. CHs are also the reason for recurrent GSs during the solar minimum [6, 9–11]. This is the main reason why turbulence-related aircraft accidents are more common in the decreasing phase of the SA.

SCs with the highest number of turbulence-related accidents are Cycle 20; in the second, it is followed by Cycles 19, 22, and 18; in the third, Cycles 21 and 24 appear as seen in **Figures 2** and **3**. The only SC without any such kind of accident is SC 23, even if it is a SC that shows extraordinary events such as unprecedented long periods of low activity, low levels of solar irradiance, SW density, particle flux, and interplanetary magnetic field (IMF), abnormally long duration of 12.6 years, and unexpected variations in the ultraviolet region, especially not a very significant change were seen in the decreasing phase [12]. This shows that the highest value of the intensity in the SC maximum and the duration of the SC are not effective parameters for aircraft accidents. Additionally, SC 23 was unique in persistent low-latitude CH activity during the last 100 years with exceptionally weak polar MFs and sizes of sunspots and their relation to SA parameters, as mentioned in [12]. This is another proof that the duration of the SC is not an effective parameter in these accidents.

The main result we can draw from all these explanations above is that the number of turbulence-related aircraft accidents has relatively little to do with the duration of the SC, the highest value of the intensity in the SC maximum, and the number of GSs. Since most aircraft accidents occur during the decreasing phase of SA, what we need to investigate is the number of GSs in the decreasing phase, depending on the increasing number of CHs. In fact, we found that aircraft accidents occur much more often in the descending phase when we did the necessary research. So, how far ahead ICAO, EASA, FAA, and other aviation organizations about this subject are? Were necessary measures taken to prevent such accidents? What measures are currently being taken by these organizations for aircraft safety? Will more comprehensive measures be taken in the future to address safety management related to this issue?

4. Aircraft safety during turbulence and safety management systems by organizations

All the licensed authorities who deal with the reports of ACI are specially educated in this field, and they depend on ICAO which is a specialized agency of the United Nations. This agency is entrusted with controlling the principles and techniques of international air navigation, planning and developing of international air transport to ensure regular growth and to provide safer transportation [6]. Some ICAO documents have also been published, particularly on turbulence such as The Transportation Safety Board of Canada (TSB) Aviation Investigation Report, Transport Canada (TS) Advisory Circular, FAA Advisory Circular, Australian Transport Safety Bureau (ATSB) Aviation Safety Bulletin/Educational Fact Sheet/Transport Safety Report and CASA, and cabin crew safety documents during turbulence are ICAO Doc 10,086 and 91.535 of Civil Aviation Safety Regulations (CASR) [13] to teach all aircraft personnel what to expect and what to do when encountering in-flight turbulence by airlines. They must know the definition of turbulence types, duration and intensity, and stages of flight such as take-off, cruise, descent, and final approach, self-management during the face of expected or unexpected turbulence, learning to use personal electronic devices during the turbulence, etc.

In the most essential sense for nonfatal accidents, there are a few important safety precautions that the passengers must take when they are exposed to turbulence such as putting the seat belt on and keeping it fastened during the flight, paying attention to the safety instructions given by the cabin crew, reading the safety information card in the seat pocket for passengers. In *non-serious* in-flight turbulence situations, the cabin crew must follow the procedures accepted by civil aviation safety authorities; for example, pilots must always control the latest weather forecasts during in-flight routes, they communicate with other nearby aircraft to check the weather conditions up ahead, and cabin crew regularly communicate with the cockpit to ensure that pilots are aware of the conditions in the cabin to avoid any injuries in the aircraft [14]. During in-flight situations and facing very severe turbulence, the following techniques must be applied by pilots for aircraft safety: keeping autopilot and autothrust on, using the quick reference handbook (QRH) turbulence penetration speed, not using the rudder to counter turbulence, using manual thrust if autothrust variations become excessive, considering descent to a lower flight level, not overreacting to temporary overspeed excursion, using aircraft carefully and considering inputs on the sidestick and taking advantage of the fly-by-wire capability to cope with turbulence in the case of disconnection with the autopilot, using autothrust to benefit from the ground speed mini function in final approach, and at the end of the flight, reporting any severe turbulence to encounter to the Maintenance with a logbook entry [15].

Modern aircraft are equipped with state-of-the-art detection systems, such as Light Detection and Ranging (LIDAR) and enhanced weather radar, to identify turbulence well ahead of the aircraft. In addition to this, artificial intelligence and machine learning provide an additional layer of data-driven insights into possible turbulence based on up-to-the-minute analysis of changing atmospheric conditions and a comprehensive view of navigating turbulent zones [16]. However, in some situations, the turbulence is not detectable by aircraft radar, or it happens abruptly, so the crew has no warning and is unable to take any avoiding action. In such situations, serious injuries in the aircraft are unavoidable, and sometimes, if the cabin crew cannot succeed in controlling the aircraft, there is nothing to do to avoid accidents resulting in death. According to this, GSs are the most important factor that will cause us to face such a situation. Considering that ICAO has started its studies on the effects of climate change and radiation levels on aircraft accidents [17], we thought it would be important to examine the effect of GSs, which have not done a comprehensive study on aircraft accidents before.

Climate change models pointed out that the number of moderate or severe air turbulence in the transatlantic flight routes in winter flights will increase soon.

Human-induced climate change is expected to increase perpendicular wind shears at in-flight routes within the atmospheric jet streams. Such a reinforcement would increase the frequency of the shear instabilities that generate CAT. In the coming decades, the prevalence of CAT may continue to increase due to climate change [6]. Aviation industries investigate GSs and the effects of climate change on ACI, but the effect of GSs that can cause bad weather conditions on Earth and turbulence in aviation was not discussed in detail. The damaging effects of GSs on flight instruments are classified as mechanical/technical and electronic accessories of aircraft, especially at higher latitudes, and the nervous system of the pilot and cabin crew had not been seriously placed in the aviation literature; and there are still more advanced applications that aviation organizations must consider safety management programs to avoid the negative effects of GSs for aircraft safety.

So far, it has been explained what measures should be taken to protect aircraft safety against the effects of turbulence. However, the following information can be given on what measures are taken in terms of safety management authorities on a larger scale [2, 18]:

1. In high windy days, controlling turbulence while flying in high-altitude and mountainous terrain.

2. Flying early in the morning or in the evening, avoiding the 1–5 pm period of higher thermal turbulence risk while flying at lower altitudes.

3. Reducing the total time spent in high radiation zones (especially higher latitudes) to protect from the radiation during a solar storm.

4. Delaying non-urgent flights until the effect of the solar storms passes.

5. Changing in-flight route depending on altitude and latitude effects.

6. Changing the altitude might cause more fuel consumption, and it is possible to compensate for this disadvantage by adjusting the flight speed according to altitude conditions.

However, there will be some difficulties while bringing these suggestions to life as aircraft flow management at the airport causes undesired delays and costs if it cannot make a certain number of take-offs and landings during a certain period due to restrictions. Reasons for this limited capability include the rapid increase in customer demand, and the difficulty of increasing the capacity of the system by establishing new airport facilities or expanding existing ones, as well as being excluded from existing analytical and simulation tools [19]. Increasing the number of flights without planning to cover these costs and surviving the competition leads to an increase in the number of aircraft accidents by employing airline personnel for more hours [20].

5. Conclusions

The weather-related effects such as ice, snow, frost, wind, turbulence, convection, pollution dust, visibility, volcanic ash, and wind shear; and during in-flight, turbulence and icing as CAT is decreased thrust and lift, increased drag and stall speed, and modified handling characteristics [21]. In addition to the well-known meteorology-related aircraft accidents, there is a strong correlation between GSs and Earth's space weather, and GA usually plays an important role in global temperature change, hence climate change [6]. The critical importance of global climate change for aviation operations is also important because aircraft aerodynamical conditions and climb rates can be affected by temperature changes only for a few centigrade degrees on the Earth and this will strongly affect the turbulence [21].

In this study, 35 turbulence-related aircraft accidents resulting in fatalities between the years January 21, 1928, and May 21, 2024, were investigated. The geographical latitudes at which these accidents occurred were found to be 40% ($\geq 40°$), 37% ($30° \leq \varphi < 40°$), and 23% ($<30°$), respectively. The time of occurrence of all these accidents coincided with the GSs. When we look at the severity of these GSs, it is seen that their distribution is moderate (45%), weak (42%), and strong/very strong (12%). Another important result of this study is that most of these accidents were found to have occurred in the decreasing phase of SACs or in the final stages of the increasing phase. It is also found that the number of turbulence-related aircraft accidents that resulted in death from higher to lower occurred in January, April, and October and in the equinoxes in March and September; and in February, July, August, and December in the second order, and November is the month that only one accident was analyzed. In addition to this, the highest number of turbulence-related aircraft accidents seen is SC 20. The other cycles that follow from higher to lower are SC 19, SC 22, SC 18, SC 21, and SC 24, respectively. The only SC without any such kind of accident is SC 23.

All the accidents that we found as a result of our research correspond to accidents that aviation authorities have still not been able to answer and that have been included in the literature as accidents with unknown causes. However, it will be seen that they coincide with the periods of GSs when these accidents are carefully analyzed. The reason why there are more accidents, especially during the decreasing phase of SA, is that the number of CHs increases during this period. The increase in the number of CHs means that the impact of high-energy particles on the Earth is much higher than in periods of CMEs. As we have explained above, aviation authorities are carrying out various studies to minimize the effects of turbulence and have even recently started to work on climate change. However, they have not yet studied how effective GSs can be or have labeled this important near-space phenomenon as having meteorological effects. Here, aviation authorities should conduct joint research with scientists working on the relevant subject of what kind of measures can be taken in such a crisis, both in terms of aircraft safety and safety management systems. Otherwise, it will not be possible to minimize this crisis with current efforts.

Acknowledgements

This chapter was started to prepare during the author's scientific collaboration on the "Simulations of Shock Wave Propagation in the Solar Corona and Comparison with Observations" at the Astronomical Institute of Czech Academy of Sciences, Department of Solar Physics, as a Sabbatical leave. The author would like to thank the anonymous referee(s) for valuable comments and guidance. This research received no specific grant from any funding agency in the public, commercial, or not-for-profit sectors.

Conflict of interest

The authors declare no conflict of interest.

Author details

Ümit Deniz Göker
Department of Unmanned Aerial Vehicles, Vocational School, Beykent University,
İstanbul, Turkey

*Address all correspondence to: udenizg@gmail.com

IntechOpen

© 2025 The Author(s). Licensee IntechOpen. This chapter is distributed under the terms of the Creative Commons Attribution License (http://creativecommons.org/licenses/by/4.0), which permits unrestricted use, distribution, and reproduction in any medium, provided the original work is properly cited. (cc) BY

References

[1] Pilot Institute. Types of Turbulence Explained. 2025. Available from: https://pilotinstitute.com/types-of-turbulence/ [Accessed: April 10, 2025]

[2] CFI Notebook. Turbulence. 2025. Available from: https://www.cfinotebook.net/notebook/weather-andatmosphere/turbulence [Accessed: April 5, 2025]

[3] Safety Management International Collaboration Group. Safety Management Terminology. SM ICG. EASA Headquarters, Cologne, Germany; 2022

[4] Plane Crash Info. 2025. Available from: https://www.planecrashinfo.com/ [Accessed: April 5, 2025]

[5] National Transportation Safety Board (NTSB). 2025. Available from: https://www.ntsb.gov/safety/data/Pages/Data_Stats.aspx [Accessed: April 5, 2025]

[6] Aksen U, Göker ÜD, Timoçin E, Akçay Ç, İpek M. The effect of geomagnetic storms on aircraft accidents between the years 1919-2023 in civil aviation. Advances in Space Research. 2024;**73**(1):807-830. DOI: 10.1016/j.asr.2023.11.008

[7] Göker ÜD. Hypothesis on the effects of geomagnetic storms on cognitive states of pilots. Defence Science Journal. 2018;**17**(2):115-138. ISSN (Online): 2148-1776

[8] Federal Aviation Administration. 2025. Available from: https://www.faa.gov/ [Accessed: October 24, 2024].

[9] Hundhausen AJ, Sime DG, Hansen RT, Hansen SF. Polar coronal holes and cosmic-ray modulation. Science, New Series. 1980;**207**(4432):761-763. DOI: 10.1126/science.207.4432.761

[10] Gushehina RT, Belov AV, Tlatov AG, Yanke VG. Coronal holes in the long-term modulation of cosmic rays. Geomagnetism and Aeronomy. 2016;**56**(3):257-263. DOI: 10.1134/S0016793216030063

[11] Venkatesan D, Agrawal SP, Lanzerotti LJ. On the three-dimensional nature of the modulation of galactic cosmic rays. Journal of Geophysical Research. 1980;**85**(A12):6893-6894. DOI: 10.1029/JA085iA12p06893

[12] Mursula K, Holappa L, Lukianova R. Seasonal solar wind speeds for the last 100 years: Unique coronal hole structures during the peak and demise of the grand modern maximum. Geophysical Research Letters. 2016;**44**(1):30-36. DOI: 10.1002/2016GL071573

[13] Australian Government, Civil Aviation Safety Authority. Crew Safety During Turbulence (Advisory Circular AC 91-28 v1.1). 2025. Available from: https://www.casa.gov.au/crew-safety-during-turbulence [Accessed: May 4, 2025]

[14] Australian Government, Australian Transport Safety Bureau. Staying Safe Against In-flight Turbulence (AR-2008-034). 2014. Available from: https://www.atsb.gov.au/sites/default/files/media/4718845/AR-2008-034Turbulence FactSheet_v2.pdf [Accessed: May 4, 2025]

[15] Graef R, Vieu JP, Spataro D. Airbus Aviation Safety Department. Operations (Managing Severe Turbulence). 2025. Available from: https://safetyfirst.airbus.

com/managingsevere-turbulence/ [Accessed: May 4, 2025]

[16] Favela R. Tommorrow.io. Mitigating Turbulence Risks: Strategies for the Aviation Industry. 2025. Available from: https://www.tomorrow.io/blog/ mitigating-turbulence-risks-aviation-strategies/ [Accessed: May 4, 2025]

[17] Williams PD. Increased light, moderate and severe clear-air turbulence in response to climate change. Advances in Atmospheric Sciences. 2017;**34**(5):576-586. DOI: 10.1007/s00376-017-6268-2

[18] Meier MM, Copeland K, Klöble KEJ, Matthiä D, Plettenberg MC, Schennetten K, et al. Radiation in the atmosphere-A hazard to aviation safety? Atmosphere. 2020;**11**:1358-1389. DOI: 10.3390/atmos11121358

[19] Lucertini M, Smriglio S, Telmon D. Network optimization. In: Dell'Olmo P, Odoni AR, editors. Air Traffic Management, Modelling and Simulation of Air Traffic Management in Lucio Bianco. Berlin: Springer; 1997. DOI: 10.1007/978-3-642-60836-0_5

[20] Calderón DJ. Aviation Investment Economic Appraisal for Airports, Air Traffic Management. Airlines and Aeronautics. England: Ashgate Publishing Limited; 2020. DOI: 10.4324/9781315178455

[21] Gultepe I, Feltz WF. Aviation meteorology: Observations and model. Introduction. Pure and Applied Geophysics. 2019;**176**(5):1863-1867. DOI: 10.1007/s00024-019-02188-2

Chapter 4

Diagnosis and Monitoring of Electrical Wiring and Interconnection Systems

Wafa Ben Hassen

Abstract

The continued operation of aircraft beyond their initially intended service life, combined with the increasing electrification of onboard systems, has intensified the need for reliable diagnosis and monitoring of electrical wiring interconnection systems (EWIS). The latter usually operates in harsh environments, exposed to mechanical, thermal, and electromagnetic stresses that can lead to faults such as insulation damage, conductor breaks, and connector failures. Moreover, since the EWIS is often embedded within the aircraft structure, it presents significant challenges for fault detection and localization, yet failures can have catastrophic consequences. In this context, the early diagnosis and monitoring of transient faults (i.e., arcing) and soft faults becomes mandatory to enable predictive maintenance strategies aimed at enhancing operational safety and reducing unscheduled downtimes and costs. Reflectometry has emerged as a non-intrusive and cost-efficient technique, enabling fault detection and localization through the analysis of signal reflections caused by impedance discontinuities. Advanced variants based on spread spectrum and multi-carrier reflectometry have been developed for real-time monitoring. However, realistic operating conditions impose limitations related to signal attenuation, distortion, and complex topologies. This chapter proposes to study reflectometry-based strategies to overcome challenges imposed by the aircraft environment, such as embedded and distributed diagnostic approaches. These strategies, leveraging multiple sensor nodes and real-time data fusion, represent a key enabling technology for next-generation intelligent EWIS health monitoring systems in safety-critical aerospace environments.

Keywords: EWIS, reflectometry, fault, diagnosis, monitoring

1. Introduction

As aircraft continue to operate well beyond their original intended service life, their electrical wiring and interconnection systems (EWIS) are increasingly exposed to harsh operational environments (i.e. mechanical, thermal, and electromagnetic stresses). These conditions contribute over time to a variety of degradations, ranging from minor insulation fissures to complete conductor breaks, that compromise the

IntechOpen

integrity and reliability of electrical networks of onboard electrical systems. Statistically, the most common faults in aging aircraft include insulation damage (37%), conductor breaks (11%), and connector failures (9%). These electrical faults are typically hidden within the aircraft structure and therefore more difficult to locate. However, their consequences can be catastrophic. The fatal accidents of TWA Flight 800 in 1996 and Swissair Flight 111 in 1998, both attributed to electrical wiring failures, underscore the critical need for continuous health monitoring of these systems [1].

Beyond aging aircraft, the complexity and vulnerability of electrical systems have increased with the global trend toward electrification. This transformation, particularly through the adoption of X-by-wire technologies (e.g., fly-by-wire and brake-by-wire), is replacing traditional mechanical and hydraulic systems with software-controlled electrical architectures. In aviation, electrical power demand has increased from 320 kW in the Airbus A320 to nearly 800 kW in the Airbus A380, accompanied by a corresponding increase in wiring length, reaching up to 530 km in some modern aircraft [2]. Although these advances enable greater functionality, efficiency, and weight reduction, they also introduce new challenges related to the reliability of EWIS.

In this context, one of the major challenges lies in the early detection and prediction of soft faults—early-stage degradations that do not yet impact functionality but may lead to arcing, short circuits, or electromagnetic interference. Soft faults, which may appear as partial insulation damage, are difficult to detect because of their low electrical signature, mainly in the presence of noise, vibration, and temperature cycling. However, their identification is critical to enabling predictive maintenance. The latter aims to maximize system availability and safety while minimizing unscheduled downtime and maintenance costs.

To meet these objectives, several diagnostic techniques have been developed in the literature, including infrared thermography, X-ray imaging, and continuity testing. However, these approaches are generally intrusive, localized, or not suitable for real-time monitoring. Reflectometry has emerged as one of the most promising methods for fault detection and localization in electrical networks. The principle of reflectometry consists of injecting a test signal into the wiring system and analyzing the reflections generated by impedance discontinuities. Time and frequency-domain variants, including spread spectrum time-domain reflectometry (SSTDR), multi-carrier time-domain reflectometry (MCTDR) [3], and multi-tone time-domain reflectometry (OMTDR) [4], have been explored to mitigate interference with native signals and enable *monitoring* during system operation.

Furthermore, the concept of *embedded diagnosis*, where the monitoring function is integrated directly into the target system, is increasingly important for real-time monitoring. However, in complex wiring architectures, using a single reflectometry sensor may be insufficient due to signal attenuation and network topology complexity. This has led to the development of *distributed diagnostic* strategies, which rely on multiple sensors placed at various endpoints of the network. Such distributed systems can improve coverage and fault location accuracy but also raise challenges related to sensor synchronization, signal interference, and data fusion.

Ultimately, the combination of embedded, distributed, and real-time monitoring approaches centered on advanced reflectometry-based techniques represents a key enabler for the next generation of intelligent and reliable EWIS monitoring in increasingly electrified and safety-critical aerospace systems.

2. The EWIS: A critical system in aging and electrified aircraft

2.1 Historical context and safety-driven evolution

The development of a regulation for EWIS in aircraft was largely motivated by two major aviation disasters: the explosion aboard TWA Flight 800 in 1996 involving a Boeing 747 and the onboard fire of Swissair Flight 111 in 1998 on a McDonnell Douglas MD-11. Both events tragically resulted in the total loss of aircraft, and all lives on board. These incidents highlighted the critical need to reassess the role of wiring in overall aircraft safety, which had previously been underestimated in comparison to other onboard systems.

In response, new regulatory standards were introduced, notably through the Federal Aviation Administration (FAA) in the United States. The requirements for EWIS certification in transport-category airplanes were formally incorporated into the Federal Aviation Regulations (FAR).

The National Transportation Safety Board (NTSB) concluded that the probable cause of the TWA 800 accident was an explosion in the center wing fuel tank. Investigators considered an electrical short circuit outside the tank—likely linked to the fuel quantity indication system wiring—as the most probable source of the accident. Similarly, the investigation into Swissair Flight 111 by the Transportation Safety Board of Canada pointed to wire arcing as the likely origin of the fire. According to the FAA, an EWIS is defined as "any wire, wiring device, or combination of these, including termination devices, installed in any area of the aircraft for the purpose of transmitting electrical energy, including data and signals" [5]. This includes not only the electrical cables themselves but also connectors, splices, clamps, backshells, bonding components, and other secondary protective elements, as shown in **Figure 1**.

The increasing reliance on electrical and electronic systems—particularly with the adoption of *X-by-wire* technologies (e.g., fly-by-wire and brake-by-wire)—has led to a sharp rise in the complexity of onboard wiring. This electrification trend improves weight efficiency and system performance but also introduces new challenges in terms of fault diagnosis and maintenance operations. As the wiring system becomes increasingly critical for both control and energy distribution, ensuring its integrity and enabling advanced diagnostic capabilities have become essential for aircraft safety and reliability.

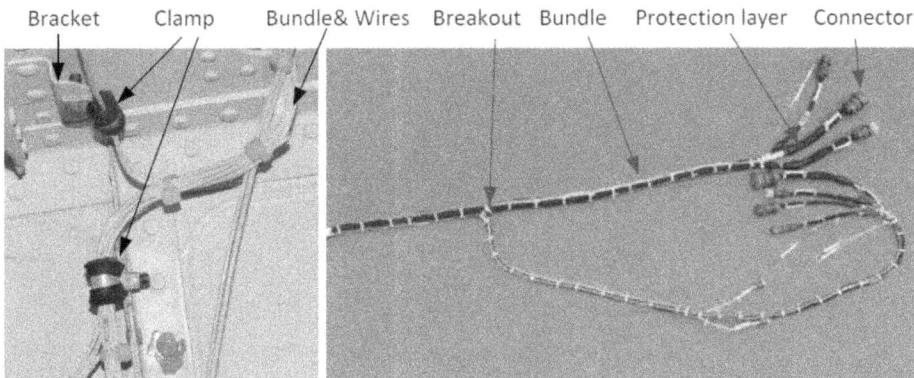

Figure 1.
EWIS in an aircraft (left) and an EWIS with an independent view (right) [6].

2.2 Degradation mechanisms and reliability issues

Over time, aircraft electrical wiring systems are subjected to a wide range of operational and environmental stresses that progressively alter their physical and electrical integrity. Aging aircraft, in particular, accumulate various forms of degradation that affect not only the cables themselves but also associated components such as connectors, splices, and protective elements. The most common degradation mechanisms include insulation chafing caused by vibrations and abrasion, thermal stress from temperature cycling, corrosion due to humidity or fluid ingress, and mechanical damage. Moreover, a significant proportion of wiring faults—up to 80%, according to NASA—are caused by human interaction, either during maintenance activities. Statistical analysis reports that the most frequent issues in EWIS involve insulation damage (37%), conductor breaks (11%), and connector malfunctions (9%) [7].

Depending on their severity and nature, wiring faults can be classified into three main categories:

- *Hard faults* refer to permanent and clearly identifiable failures such as open circuits or short circuits. These faults result in the complete interruption of power or signal transmission and typically require reactive maintenance.

- *Soft faults* correspond to incipient or partial degradations that cause weak variations in cable impedance. They are often linked to insulation deterioration, partial conductor thinning, or micro-cracks, as shown in **Figure 2**. Although soft faults do not directly interrupt EWIS functionality, they can lead to increased local heating and signal distortion and eventually evolve into hard faults (**Figure 3**). Such degradations are further illustrated in **Figure 4**, which presents typical examples observed in aircraft: cracks in the cable sheath, fraying of insulation due to mechanical stress, and partial insulation removal, potentially exposing conductors to arcing and electromagnetic interference.

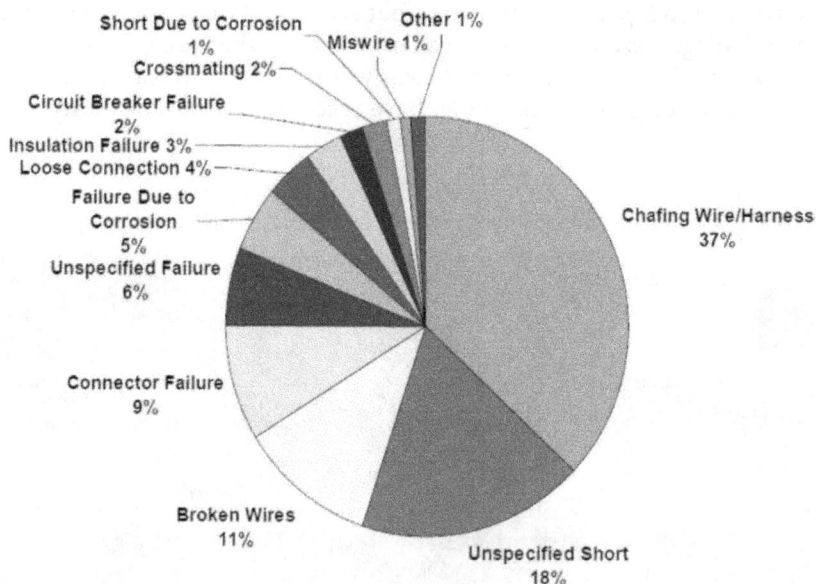

Figure 2.
Statistical breakdown of cable faults in aircraft based on NASA's 2007 data [6].

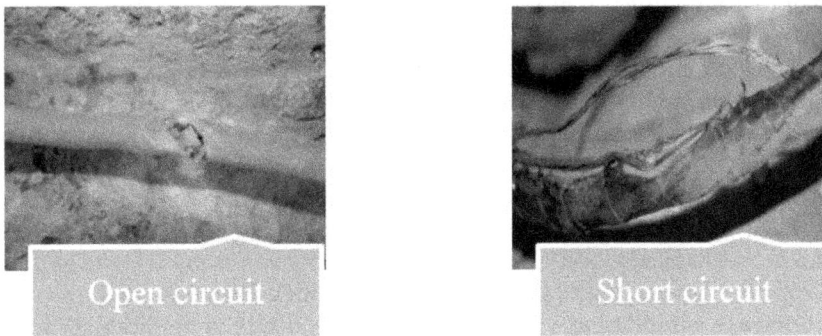

Figure 3.
Examples of hard faults: open circuit (left) and short circuit (right).

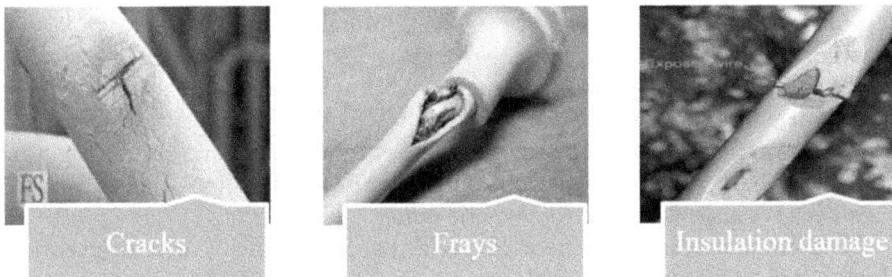

Figure 4.
Examples of soft faults: cracks in protective layers (left), frayed insulation exposing conductors (middle), and partial insulation damage with visible exposed wire (right).

- *Intermittent faults*, including arcing events, represent a particularly hazardous class of failures. These faults appear sporadically, often triggered by environmental conditions such as vibration, humidity, or electrical transients. They can cause high-energy discharges, generate carbon tracking, and even initiate fires—posing significant safety risks.

As aircraft fleets age and electrification continue to be attractive, the dependency on EWIS is also increasing. The convergence of physical degradation mechanisms, increased system complexity, and the growing reliance on electrical functions call for proactive fault prevention strategies and a deeper understanding of fault impact in real-world operational conditions.

3. Diagnostic and monitoring techniques for EWIS integrity

In the field of aircraft electrical wiring diagnostics, a wide range of techniques have been explored to detect, locate, and characterize faults in complex embedded systems. Among the most established methods are visual inspection, X-ray imaging, capacitive and inductive measurements, optical and acoustic sensing, and reflectometry. Visual inspection, though widely used in maintenance routines, offers limited effectiveness —particularly in aircraft where a large portion of the wiring is concealed within structural components or densely packed harnesses. According to NASA, this method detects no more than 25% of wiring faults in aircraft EWIS [7]. X-ray imaging, while capable of revealing internal or hidden anomalies, requires heavy equipment and full cable access, making it impractical for routine or embedded applications.

Capacitive and inductive methods permit the detection of faults in point-to-point cable configurations but quickly lose effectiveness in multi-branched networks. Additionally, they require system downtime for testing, which limits their usability for real-time monitoring.

Beyond these conventional approaches, recent research has explored alternative techniques such as acoustic and optical sensing. Acoustic methods, based on the detection of mechanical vibrations or acoustic emissions caused by faults or arcing events, can offer intermittent fault detection capabilities. However, their sensitivity to ambient noise and limitations in fault localization reduce their applicability in the aircraft domain. Optical sensing technologies—particularly those based on fiber Bragg gratings (FBG) or optical time-domain reflectometry (OTDR)—provide distributed, high-resolution monitoring of thermal and mechanical faults along cables. While promising for embedded and real-time monitoring, these systems require specialized optical infrastructure, making retrofitting in existing fleets highly expensive.

Another emerging class of techniques involves electromagnetic signature analysis (EMSA), which captures anomalies in the electromagnetic emissions generated by damaged wiring. These methods are particularly effective for detecting intermittent faults, such as arcing, and can operate passively without disturbing system function. Nevertheless, their ability to accurately localize faults remains limited, and their performance is sensitive to electromagnetic noise.

Table 1 summarizes the comparative performance of some of these diagnostic methods based on several key criteria relevant to aircraft EWIS fault scenarios.

Reflectometry, particularly in its time-domain form (TDR), consistently emerges as one of the most robust and versatile solutions for aircraft wiring diagnostics. It not only demonstrates strong capabilities across various fault types—including hard, soft, and intermittent faults—but also provides precise fault localization. Its non-intrusive nature, which allows test signals to coexist with native ones, makes it particularly attractive for online and embedded applications. Unlike many emerging technologies, it has reached a high Technology Readiness Level (TRL) and is already deployed in operational contexts

Scenario/Criterion	Visual	X-Ray	Cap./Ind.	FDR	TDR	Acoustic	Optical
Long cable	−	−	+	++	++	+	++
Buried cable	−	−	+	++	++	−	++
Soft fault	+	++	−	+	+	−	+
Intermittent fault	−	−	−	+	++	+	+
Online diagnosis	−	−	−	+	++	+	++
Multi-branched network	−	−	−	−	+	−	+
Cost-effectiveness	++	−	++	+	+	+	−
Fault localization accuracy	−	+	−	++	++	−	+
Weight impact	++	−	++	+	+	+	−

(++): highly effective or favorable; (+): conditionally effective or moderate; (−): not effective or not applicable. Cap./Ind. = Capacitive and Inductive methods; FDR = Frequency-Domain Reflectometry; TDR = Time-Domain Reflectometry.

Table 1.
Assessment of diagnostic methods for detecting wiring faults in aircraft.

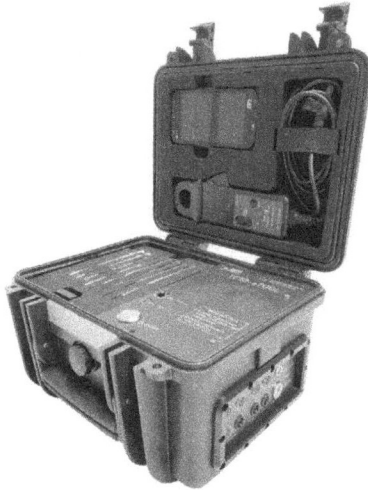

Figure 5.
Integration of OMTDR into a portable electrical troubleshooting test tool, the Electrical Ground Support Equipment (E-GSE) TC50 e-tool.

Figure 6.
Integration of MCTDR into a portable tool, Aero Smart-R kit.

such as Maintenance, Repair, and Overhaul (MRO). **Figures 5** and **6** present examples of reflectometry-based technology transfers to the aircraft industry, such as OMTDR and MCTDR, respectively.

For monitoring purposes, reflectometry-based technologies are well-suited due to their moderate weight impact and adaptability to distributed architectures, which make them scalable to complex aircraft wiring networks. These combined attributes make reflectometry a key enabling technology for both current and next-generation EWIS monitoring systems.

4. Reflectometry-based faults monitoring and diagnostic in EWIS

Electrical reflectometry has emerged as one of the most promising and versatile techniques for detecting and localizing faults in the EWIS. It is based on the principle that any discontinuity in the electrical impedance of a transmission line—caused, for instance, by insulation degradation, connector issues, or a broken conductor—will reflect part of an injected signal back to its origin. By analyzing these reflected signals, known as echo responses, it is possible to determine the presence, type, and location of faults along the system under test.

This method is non-intrusive, compatible with real-time monitoring, and particularly well-suited for embedded systems such as aircraft EWIS, where direct physical access to wiring is often impossible. Reflectometry supports both offline and online diagnostics. Combined with advanced processing and intelligence techniques, it can be integrated into predictive maintenance architectures to improve system safety and reduce unscheduled downtime. Aircraft on-ground situations can be extremely costly, with some airlines estimating losses of up to 150,000 USD per hour. For instance, in 2011, a significant incident occurred when a crack was discovered in the fuselage of a Boeing 737, leading to the cancelation of approximately 300 flights. As a precaution, 79 aircraft of the same type were grounded until inspections and repairs could be completed. This disruption reportedly resulted in financial losses reaching 4 million USD due to halted operations and lost revenue [1].

Reflectometry techniques are commonly divided into two main families: Time-Domain Reflectometry (TDR) and Frequency-Domain Reflectometry (FDR). In TDR, a fast electrical pulse or voltage step is injected into the cable. When this signal encounters a change in impedance, part of it is reflected back to the source. The reflected signal, called a reflectogram [8], contains echoes of the original pulse. By measuring the time delay between the emitted and reflected signals and knowing the propagation velocity of the signal in the cable, the distance to each fault can be accurately calculated.

FDR, in contrast, operates in the frequency domain. It involves injecting a continuous or swept sinusoidal signal (chirp) and analyzing the frequency response of the line. Impedance discontinuities cause standing wave patterns, from which information about their location and their nature (i.e., hard/soft fault) can be extracted. While FDR can be more sensitive to soft fault, its analysis becomes more complex in multi-branched networks, making interpretation more challenging. TDR is often preferred in aviation applications, especially in complex EWIS architectures due to its suitability for onboard integration (i.e., electronic board and chip) and its localization accuracy. In this context, several diagnostic strategies have been developed, including advanced methods such as SSTDR, OMTDR, and MCTDR, which are specifically designed to enable fault monitoring in increasingly electrified and distributed aircraft systems.

4.1 Fundamental principles of electrical reflectometry

Electrical reflectometry relies on modeling the cable under test as a distributed parameter system, also known as a transmission line. This representation takes into account the four per-unit-length electrical parameters: resistance R, inductance L, capacitance C, and conductance G, which influence how voltage and current propagate along the line. The evolution of voltage $v(z,t)$ and current $i(z,t)$ along the cable is obtained by the resolution of Telegrapher's equations.

From these equations, two fundamental transmission line parameters are derived. First, the characteristic impedance Z_c, which relates voltage to current in steady-state sinusoidal conditions:

$$Z_c = \sqrt{\frac{R + j2\pi f L}{G + j2\pi f C}}.$$ (1)

Second, the propagation constant γ, which accounts for both signal attenuation and phase shift as the wave travels along the transmission line:

$$\gamma = \sqrt{(R + j2\pi f L)(G + j2\pi f C)}. \tag{2}$$

When an electrical signal encounters an impedance discontinuity in the line—such as a connector fault, insulation degradation, or a conductor break—a portion of the signal is reflected. The magnitude and polarity of this reflection are described by reflection coefficients. At the injection point, the reflection coefficient Γ_E at the system input (at the interface between the generator and the transmission line) is defined as:

$$\Gamma_E = \frac{Z_c - Z_0}{Z_c + Z_0}. \tag{3}$$

At the end of the line, the load reflection coefficient Γ_L, is given by:

$$\Gamma_L = \frac{Z_L - Z_c}{Z_L + Z_c}. \tag{4}$$

The overall frequency response of the cable can be described by the transfer function $H(f)$, which models the superposition of multiple reflected signals:

$$H(f) = \Gamma_E + \frac{(1 - \Gamma_E^2)\Gamma_L e^{-2\gamma l}}{1 + \Gamma_E^2 \Gamma_L e^{-2\gamma l}}. \tag{5}$$

where l is the distance to the fault, and γ is the complex propagation constant introduced in Eq. (4).

In TDR, a short pulse is injected into the cable, and the reflected signal—known as the reflectogram—is analyzed. The distance to the fault d is computed from the time delay Δt of the echo using:

$$d = \frac{v\Delta t}{2}, \tag{6}$$

where v is the signal propagation velocity in the cable.

4.2 Soft fault diagnosis: Challenges and proposed solutions

In electrical reflectometry, especially with TDR, the ability to detect and localize soft faults is a critical challenge. These faults refer to minor degradations or structural deviations along the cable that do not completely disturb the signal but instead create weak impedance discontinuities. In contrast to hard faults (e.g., open or short circuits), soft faults generally produce partial signal reflections whose amplitudes vary with fault severity, making them difficult to distinguish from noise, as shown in **Figure 7**.

On the reflectogram, a spread soft fault appears as a low-amplitude deviation, generally characterized by two reflection points: one at the entry and one at the exit of the spreading fault with length L_d. To ensure effective detection and localization accuracy of such faults, the injected signal must have sufficient bandwidth.

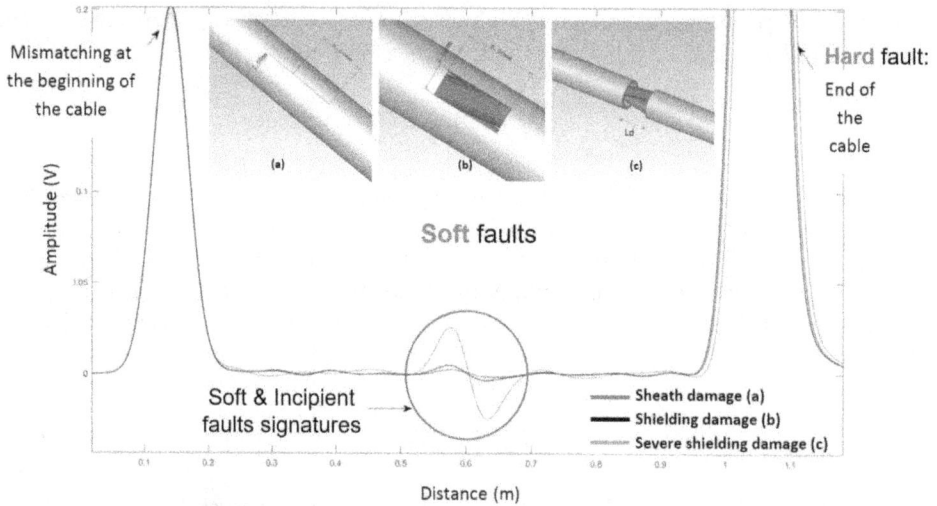

Figure 7.
Reflectometry response to insulation degradation in MLB Twisted pair used in aircraft industry.

As demonstrated in [9], there is a fundamental limit on the spatial resolution that can be achieved with a given maximum frequency f_{max} of the test signal. A fault can only be resolved if its physical length L_d satisfies:

$$L_d \gtrsim \frac{v}{4f_{max}}, \tag{7}$$

where v is the signal propagation velocity along the cable. Eq. (7) highlights that finer spatial resolution (i.e., the ability to detect shorter faults) requires increasing the bandwidth of the excitation signal. However, increasing f_{max} leads to greater signal attenuation and dispersion, which in turn reduces both the diagnosis coverage and the accuracy of fault localization. Therefore, a trade-off must be found between achieving high resolution and minimizing signal attenuation along the cable. In practice, this often means that only faults longer than a certain threshold can be meaningfully detected and characterized using reflectometry.

However, in real-world aircraft applications, EWIS is subjected to dynamic stresses—mechanical vibrations, electromagnetic interference, and thermal cycling—that may alter the echoes signatures that become highly distorted and difficult to interpret with precision. In [10], a comparative study of TDR peak responses for different fault types highlights distinct variations in detectability. The results show that signal peaks generated by minor insulation damage or superficial fraying are often comparable in amplitude to background noise, rendering their detection and localization highly unreliable in real-world settings. Conversely, open and short-circuit faults produce much stronger reflections, with peak amplitudes that are clearly distinguishable and easily identified using TDR.

These findings underline the inherent limitations of TDR when applied to soft faults such as early-stage insulation degradation or small radial cracks. The minimal impedance changes they introduce result in weak electrical signatures that can be easily masked by noise. This emphasizes the importance of developing enhanced diagnostic techniques to improve the sensitivity and robustness of reflectometry-based methods, particularly in the context of complex and noise-prone wiring environments.

To overcome these limitations, various signal-processing strategies have been developed in the literature to enhance the detection of soft fault signatures [11–14]. Among them, the Self-Adaptive Correlation Method (SACM) [15] and the Signature Magnification by Selective Windowing (SMSW) [16], aim to amplify weak reflections through adaptive windowing and correlation techniques. Probabilistic approaches incorporating baseline measurements and data fusion have also been proposed to improve robustness in noisy conditions [4]. Optimization algorithms such as genetic algorithms have been used to extract and estimate physical parameters related to soft faults, including length, inductance, or capacitance [8, 17]. Most of these approaches, however, rely on the availability of a correct reference signal from a healthy system state and often assume simplified lossless models that are valid only under laboratory conditions. In practice, this limits their performance in complex wiring environments such as those encountered in aging aircraft. To address this, the NASA research team has explored physics-based modeling approaches for early-stage fault simulation in specific cable types (e.g., coaxial and twisted pairs) [18, 19]. While accurate, these models are highly complex, computationally intensive, and not generalizable to large-scale embedded systems.

More broadly, the issue of soft fault detection has been addressed in other engineering domains—particularly rotating machines—using advanced signal processing methods such as spectral kurtosis [20, 21], Principal Component Analysis (PCA) [22], Curvilinear Component Analysis [23], the Hilbert-Huang transform, wavelet transforms, and empirical mode decomposition [11]. These tools, often combined with neural networks [24] or genetic algorithms, have demonstrated effectiveness in weak signal detection. Hybrid approaches, which combine physical modeling with data-driven learning, are increasingly explored for soft fault diagnosis in complex systems. These methods leverage physics-informed constraints to improve generalization and reduce the need for large datasets while maintaining the flexibility of AI-based models. Overall, the integration of reflectometry with modern signal processing, statistical inference, and hybrid modeling holds significant promise for the reliable detection of soft faults in real-world applications. In this context, an AI-assisted approach combining TDR, subtractive correlation method, and neural networks has been proposed for soft fault localization in complex aircraft microgrids, showing improved accuracy and robustness even under noisy conditions [25].

4.3 Monitoring of transient faults: Challenges and proposed solutions

Integrating a reflectometry-based system directly into the EWIS allows for monitoring of wiring integrity during system operation. This approach is particularly effective for identifying transient events, such as arcing faults, which typically occur over very short durations (on the order of 1 ms). These faults are frequently observed in avionics during flight but are difficult to reproduce under static maintenance conditions. Despite its advantages, online diagnostics presents several technical challenges. The diagnostic process must be robust against electromagnetic noise and non-intrusive. It must not interfere with native signals (i.e., communication and power signals). To address this constraint, excitation reflectometry signals should be injected in frequency bands that are either separate from or wider than those already used by the operational system.

This has led to the development of advanced reflectometry techniques such as SSTDR (2006), MCTDR (2008), and OMTDR (2013). In the case of an OMTDR signal based on Orthogonal Frequency Division Multiplexing (OFDM), each subcarrier is associated with a complex coefficient s_k, which encodes both the amplitude and phase of the modulated symbol—typically from a Binary Phase Shift Keying (BPSK), Quadrature Phase Shift Keying (QPSK), or Quadrature Amplitude Modulation (QAM) constellation. The time-domain representation of the excitation signal is obtained by summing all subcarriers as follows:

$$x[n] = \sum_{k=0}^{N-1} s_k \cdot e^{j\frac{2\pi k}{N}n}, \quad n = 0, 1, \dots N-1 \tag{8}$$

In this equation, $x[n]$ denotes the nth time-domain sample of the OFDM signal, N is the total number of subcarriers, and s_k is the complex spectral coefficient (or modulated symbol) assigned to the kth subcarrier. The exponential term $e^{j\frac{2\pi k}{N}n}$ represents the kth sinusoidal basis function at frequency $f_k = k \cdot \Delta f$, where Δf is the subcarrier spacing. The set of symbols $\{s_k\}$ forms the OFDM frame in the frequency domain.

In practical implementations, the time-domain signal is generated by computing the inverse fast Fourier transform (iFFT) of the frequency-domain vector as follows:

$$x = \text{iFFT}(\{s_k\}) \quad \text{this is,} \quad x[n] = \frac{1}{N}\sum_{k=0}^{N-1} s_k \cdot e^{j\frac{2\pi k}{N}n} \tag{9}$$

This transformation ensures that the time-domain signal maintains orthogonality between subcarriers and provides a well-controlled spectral structure, which is crucial for reflectometry-based diagnostics in constrained environments.

To enable the use of the multi-carrier method for monitoring, the test signal must be carefully designed to comply with frequency constraints imposed by the system's electromagnetic environment or regulatory standards. In particular, certain frequency bands may be prohibited to avoid interference with other onboard equipment. As a result, the frequency components of the signal corresponding to these prohibited bands must be eliminated. This constraint can be expressed by the following condition applied to the spectral coefficients s_k of the excitation signal:

$$s_k = 0 \quad \text{for all} f_k = k \cdot \Delta f \in \mathscr{F}_{\text{forbidden}}, \quad k \in \{0, \dots, N-1\} \tag{10}$$

In this expression, s_k represents the complex spectral coefficient associated with the kth subcarrier of the excitation signal. Each subcarrier operates at a frequency $f_k = k \cdot \Delta f$, where Δf denotes the frequency spacing between adjacent subcarriers, commonly referred to as the subcarrier step. The set $\mathscr{F}_{\text{forbidden}}$ includes all frequencies that must be excluded from the excitation signal. The total number of subcarriers is denoted by N, which defines the bandwidth of the OMTDR excitation signal. This condition ensures that the excitation signal does not include energy in prohibited spectral bands, which is essential for compliance with electromagnetic compatibility (EMC) constraints or system-level interference avoidance.

However, this results in a significant loss of information in the reflectometry signal. **Figure 8** illustrates the impact of cancelation-specific subcarriers—particularly those corresponding to frequency bands that must be avoided for system compatibility reasons. It clearly shows that the distortion around the main peak (named artifacts) increases as the width of the forbidden frequency band grows.

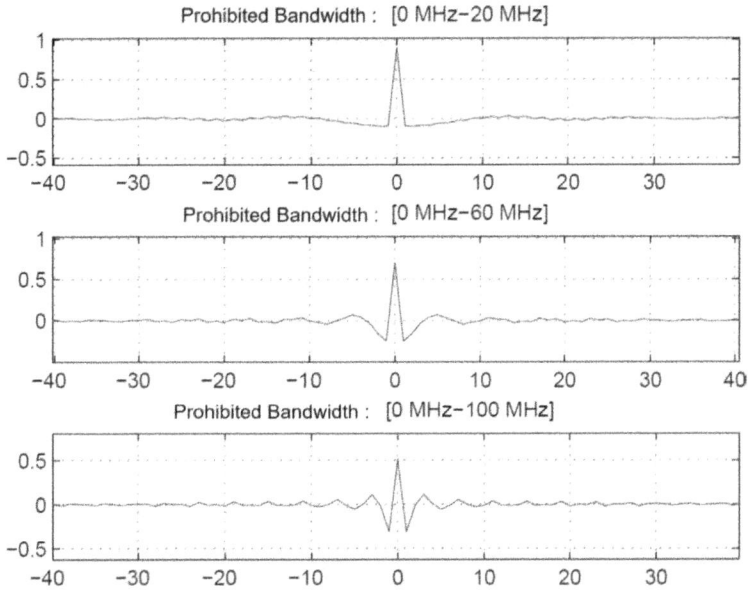

Figure 8.
Influence of subcarriers cancelation on prohibited frequency bands: Presence of distortion around the main peak.

To address the information loss caused by the putting to zero of forbidden frequencies, a post-processing module based on the CLEAN algorithm has been introduced [3]. The CLEAN algorithm operates iteratively by identifying peaks in the reflectogram in descending order of amplitude (in absolute value) and successively removing their contributions. The process continues until one of two stopping criteria is met: either a predefined residual error threshold is reached, or a maximum number of peaks have been extracted. However, uncertainty in peak amplitude—due to overlap between adjacent peaks—can lead to localization errors, which accumulate over successive iterations. **Figure 9** shows the reflectogram obtained using the OMTDR method on an electrical wiring network with a Y-topology composed of three branches: B_1, B_2, and B_3, with respective lengths of 6 m, 11 m, and 5 m. A soft fault is introduced on branch B_3, located 9 m from the injection point, and characterized by a 40% variation in characteristic impedance. The amplitude is normalized with respect to the main peak of the autocorrelation of the excitation signal. A deconvolution using the CLEAN algorithm (red curve) has been applied to the measured reflectogram (blue curve). The peaks detected by CLEAN are located at 6 m, 10.5 m, 11 m, 17 m, 21.5 m, and 22 m. Notably, the soft fault at 9 m was not detected by the algorithm.

When prior knowledge of the network topology is available, the number of peaks to be detected can be limited accordingly. However, this condition becomes problematic for soft faults, which are characterized by low-amplitude signatures. In complex networks, where multiple junctions, terminations, reflections, and faults coexist, the reflectogram can become highly complex. As a result, the CLEAN algorithm may lose its effectiveness in accurately detecting and localizing soft faults. To mitigate these challenges, an averaging process over several OFDM symbols can be applied to recover lost information and enhance the signal-to-noise ratio (SNR). This temporal averaging helps strengthen persistent features in the reflectogram while suppressing random noise and artifacts. Finally, a dedicated post-processing step—such as convolution with a Chebyshev window [26]—is applied to enhance peak sharpness and

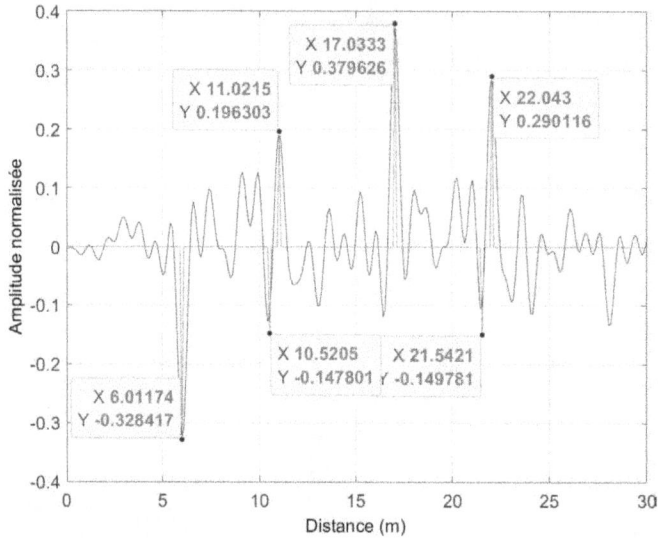

Figure 9.
Post-processed OMTDR-based reflectogram using clean algorithm.

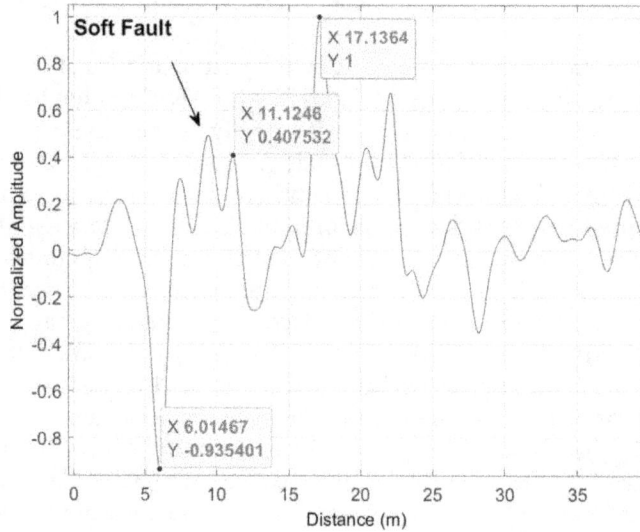

Figure 10.
Post-processed OMTDR-based reflectogram using Chebyshev window convolution.

suppress side lobes, improving the clarity and resolution of fault localization, particularly in scenarios with closely spaced reflections as shown in **Figure 10**.

5. Multi-branched wiring networks: Challenges and proposed solutions

In modern aircraft, branched electrical wiring topologies, such as star-shaped, Y-shaped, or tree-shaped configurations, are widely adopted to support the increasing complexity and distribution of onboard electrical systems. These topologies offer several advantages, including reduced cable weight, routing flexibility, and fault isolation, which are critical in aeronautical applications where space, weight, and reliability constraints are stringent. In complex wiring architectures, relying on a single injection point becomes insufficient to effectively monitor the entire network. This limitation is primarily due to the signal attenuation caused by the presence of multiple junctions. Although it is often possible to estimate the distance from the injection point to a fault, identifying the faulty branch remains challenging and ambiguous.

Although reflectometry techniques have proven to be highly effective in fault detection and localization for simple point-to-point configurations, they face significant challenges when applied to complex branched networks. The signal degradation introduced by multiple reflections and lossy paths compromises the accuracy of the diagnosis. As a result, a single-sensor setup cannot ensure full coverage of the network or distinguish between multiple possible fault localizations.

5.1 Distributed reflectometry: Optimization of resources and sensors

To overcome these limitations, a distributed diagnostic strategy is proposed. By deploying multiple sensors at various access points throughout the network, the coverage area is expanded, and ambiguity in fault localization is significantly reduced. However, this architecture introduces new challenges related to resource sharing, interference mitigation, and sensor synchronization. When several sensors are active simultaneously, their signals may interfere, requiring advanced signal processing techniques to isolate and interpret each response accurately [26]. A solution, several multiple access strategies can be implemented in the context of distributed reflectometry systems.

First, time division multiple access (TDMA) allocates a distinct time slot to each sensor for signal injection and reception. This method is straightforward to implement and avoids signal collisions. However, it is suboptimal for the detection of *intermittent faults*, as such faults may not occur during the time slot allocated to a sensor. Second, code division multiple access (CDMA) assigns an orthogonal spreading code to each sensor, allowing all sensors to operate simultaneously across the entire frequency band. This technique offers high robustness against interference and is particularly promising in scenarios involving multiple sensor networks and dynamic fault conditions. However, it requires precise synchronization and code orthogonality to avoid *cross-correlation artifacts*, which can complicate implementation. Finally, frequency division multiple access (FDMA) assigns a specific frequency sub-band to each sensor for reflectometry signal transmission and reception. This approach enables *simultaneous sensor operation*, which is more suitable for capturing transient or intermittent events. However, the division of the frequency spectrum among multiple sensors requires a *subcarrier allocation algorithm* and can lead to a reduction in reflectogram quality due to decreased bandwidth per sensor, potentially causing *information loss* and the appearance of *spectral artifacts*, as previously discussed. As a solution, it has been proposed a *distributed subcarrier allocation* method to maintain equivalent detection performance across all sensing nodes. In the example illustrated in **Figure 11**, subcarriers are alternately assigned to one of the three reflectometers, denoted as S_1, S_2, and S_3. This allocation strategy ensures that each sensor generates a multi-carrier signal composed of frequencies that are uniformly distributed across the operational bandwidth. As a result, the spectral profiles of the signals generated by S_1, S_2, and S_3 remain closely aligned, thereby guaranteeing the production of homogeneous reflectograms.

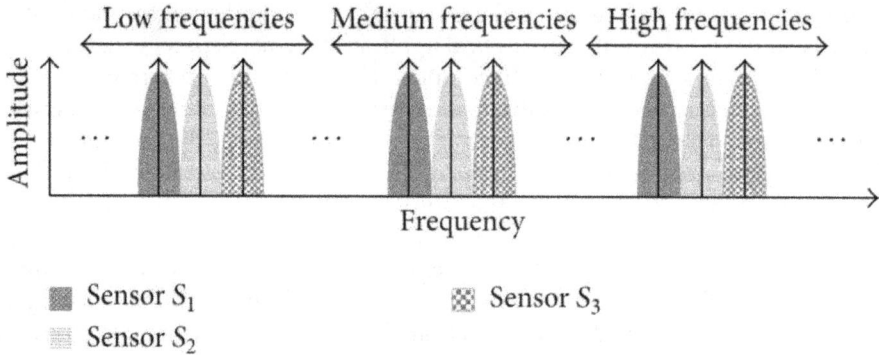

Figure 11.
Example of distributed subcarriers allocation: Three sensors implemented in the Y-shape network.

5.2 Sensors communication: Sensors fusion and BER indicator

In distributed reflectometry, enabling communication between sensors for information exchange (e.g., reflectograms and fault localizations) enhances both diagnostic coverage and reliability. To achieve this, a communication protocol—such as a *master/slave* architecture—must be defined to centralize the gathered information and facilitate fault detection and localization decisions. The *OMTDR* method, which embeds information within the transmitted test signal, allows for data transmission by utilizing the forward-propagating part of the reflectometry signal. In complex wiring networks, a *sensor clustering strategy* can be adopted, grouping sensors based on their relative positions and the network complexity (e.g., number of junctions to traverse and branch lengths) [14, 26]. Such a strategy optimizes data fusion, improves measurement consistency, and ultimately increases the accuracy and robustness of fault diagnostics in complex wiring environments.

The integration of communication between sensors has revealed a further indicator about the health state of the network. It is based on the calculation of the Bit Error Rate (BER), a key parameter used to characterize the quality of the communication channel. **Figure 12** shows the variation of the BER as a function of the soft fault characteristic impedance variation for different values of the SNR. As observed, the BER increases with the severity of the soft fault for a given SNR. Referring to **Figure 12**, the BER(S_1, S_3) corresponds to a healthy link between S_1 and S_3 ($\Delta Z_c = 0$) and BER(S_1, S_2) corresponds to a faulty link between S_1 and S_2 ($\Delta Z_c = 50\%$) in a Y-shaped network.

6. Overview of the reflectometry-based strategies for EWIS

Table 2 provides a synthetic overview of the main diagnostic and monitoring functionalities expected from EWIS systems, alongside the corresponding technical challenges and the solutions proposed in the literature. It highlights how the complexity of aircraft wiring—especially in the context of aging fleets and increased electrification—requires a combination of advanced signal processing, intelligent sensing strategies, and cost-effective deployment architectures. From the detection of

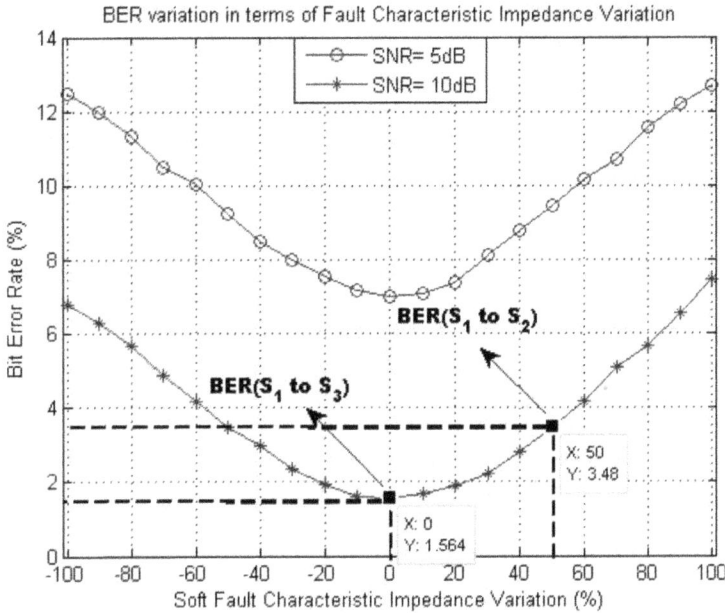

Figure 12.
Sensors Communication: the BER, an additional indicator about the health state of the EWIS.

Functionalities	Challenges	Proposed strategies
Soft fault diagnosis and monitoring	Low-amplitude reflections rapidly masked by noise; resolution attenuation trade-off; reflectogram distortion in high frequencies	Advanced post-processing methods (i.e., SACM, SMSW, and wavelet), data fusion, AI-based techniques, AI-hybrid models, etc.
Transient fault monitoring	Detection of faults occurring in under 1 ms; signal non-intrusiveness; compliance with EMC constraints	SSTDR, MCTDR, OMTDR, CLEAN post-processing, forbidden band mitigation, averaging, filtering, etc.
Multi-branched networks	Signal attenuation through junctions; fault ambiguity localization; Interference	Distributed reflectometry, multiple access techniques, subcarrier allocation, master/slave protocol communication, sensor clustering, sub-network definition, BER calculation, etc.

Table 2.
EWIS diagnosis and monitoring: functionalities, challenges, and proposed strategies.

soft or transient faults to the management of distributed sensing in branched networks, each functionality faces specific challenges that must be addressed through tailored methodologies. This summary illustrates the diversity and complementarity of approaches currently explored to ensure robust and efficient health monitoring of embedded wiring systems.

7. Conclusions

This study aimed to present innovative embedded diagnostic strategies for complex wired networks under operational constraints, focusing on performance, cost, and reliability. Reflectometry was selected as a core technology due to its natural integration capabilities within embedded systems. The challenges related to the harsh and complex environment of aging and increasingly electrified aircraft have been presented, and the corresponding solutions have been proposed. The first is the multi-carrier reflectometry method (i.e., OMTDR), which provides improved spectral efficiency and enables accurate fault monitoring. Advanced post-processing methods have been presented to address constraints related to information loss in the reflectogram and to enhance soft fault monitoring and characterization (e.g., optimization algorithms). The second approach is a distributed diagnostic strategy that helps overcome signal attenuation and fault localization ambiguities in complex wiring systems. To achieve this, several techniques have been deployed, such as multiple access, sensor communication, and data fusion. These contributions pave the way for more intelligent, reliable, and cost-effective health monitoring systems, making reflectometry a promising candidate for next-generation EWIS diagnostics in safety-critical aerospace applications. The application of reflectometry has also been extended beyond traditional cable fault detection to include other advanced functionalities, such as monitoring the onset of fire in aircraft systems [27] and assessing the operational health of energy storage systems [9]. In these contexts, reflectometry offers a non-intrusive, real-time monitoring capability that enhances safety and reliability by detecting early signs of anomalies or degradation. However, the TRL of electrical reflectometry is not at the same level as that reached for electrical wiring systems. The strategies proposed in this chapter may be of interest for advancing the technology to a higher level of maturity.

Abbreviations

BER	bit error rate
BPSK	binary phase shift keying
CDMA	code division multiple access
EMC	electromagnetic compatibility
EMSA	electromagnetic signature analysis
EWIS	electrical wiring interconnection system
FAA	federal aviation administration
FAR	federal aviation regulations
FBG	fiber Bragg grating
FDMA	frequency division multiple access
FDR	frequency-domain reflectometry
iFFT	inverse Fast Fourier Transform
MCTDR	multi-carrier time-domain reflectometry
NTSB	national transportation safety board
OFDM	orthogonal frequency division multiplexing
OMTDR	orthogonal multi-tone time-domain reflectometry
OTDR	optical time-domain reflectometry
PCA	principal component analysis
QAM	quadrature amplitude modulation
QPSK	quadrature phase shift keying
SACM	self-adaptive correlation method
SMSW	signature magnification by selective windowing
SNR	signal-to-noise ratio

SSTDR	spread spectrum time-domain reflectometry
TDR	time-domain reflectometry
TDMA	time division multiple access
TRL	technology readiness level

Author details

Wafa Ben Hassen
Université Paris-Saclay, CEA, List, Palaiseau, France

*Address all correspondence to: wafa.benhassen@cea.fr

IntechOpen

© 2025 The Author(s). Licensee IntechOpen. This chapter is distributed under the terms of the Creative Commons Attribution License (http://creativecommons.org/licenses/by/4.0), which permits unrestricted use, distribution, and reproduction in any medium, provided the original work is properly cited. (cc) BY

References

[1] Auzanneau F. Wire troubleshooting and diagnosis: Review and perspectives. Progress In Electromagnetics Research B. 2013;**49**:253-279. DOI: 10.2528/PIERB12100308

[2] Arrieta A, Krueger H, Singhose W. Wiring Systems in More Electric Aircraft: Current trends and future challenges. IEEE Aerospace and Electronic Systems Magazine. 2014;**29**(11):28-36. DOI: 10.1109/MAES.2014.6978492

[3] Lelong A, Carrion MO. On-line wire diagnosis using multicarrier time domain reflectometry for fault location. In: IEEE Sensors 2009. New York: IEEE; 2009. pp. 751-754. DOI: 10.1109/ICSENS.2009.5398204

[4] Ben Hassen W, Zanchetta A, Morel F, et al. Embedded OMTDR sensor for small soft fault location on aging aircraft wiring. Procedia Engineering. 2016;**168**: 364-367

[5] Federal Aviation Administration (FAA). 14 CFR Part 25.1701 – Electrical Wiring Interconnection System (EWIS). Federal Aviation Regulations; 2007. Available from: https://www.ecfr.gov/current/title-14/chapter-I/subchapter-C/part-25/subpart-H/section-25.1701

[6] Collins Aerospace. EWIS (Electrical Wiring Interconnection Systems) Harness Examples; 2023. Available from: https://www.collinsaerospace.com/what-we-do/Products-and-Services/Cabling-Systems

[7] Wheeler KR, Timucin DA, Twombly IX, Goebel KF, Wysocki PF. Aging Aircraft Wiring Fault Detection Survey. Technical Report V.1.0. Moffett Field (CA): NASA Ames Research Center; 2007

[8] Ben Hassen W, Kafal M. Shielding damage characterization in twisted pair cables using OMTDR-based reflectometry and inverse problems. In: PIERS Proceedings. IEEE; 2019. pp. 1-4

[9] Ben Hassen W, Slimani M. Lithium-ion battery tab welding diagnosis using electrical reflectometry. In: Proceedings of the 2024 IEEE AUTOTESTCON; 2024 Sep 9–12; National Harbor, MD, USA. New York: IEEE; 2024. pp. 1-6

[10] Griffiths LA, Parakh R, Furse C, et al. The invisible fray: A critical analysis of the use of reflectometry for fray location. IEEE Sensors Journal. 2006;**6**(3):697-706

[11] Sallem S, Ravot N. Joint self-adaptive correlation method and modified empirical mode decomposition for soft defect detection in cable by reflectometry. PIERS Letters. 2015;**5**(1): 59-65

[12] Smail MK, Hacib T, Pichon L, Loete F. Detection and location of defects in wiring networks using time-domain reflectometry and neural networks. IEEE Transactions on Magnetics. 2011;**47**(5):1038-1041

[13] Zhang J, Zhang Y. Analysis of time-domain reflectometry combined with wavelet transform for fault detection in aircraft shielded cables. IEEE Sensors Journal. 2016;**16**(4):1034-1041

[14] Franchet M, Ravot N, Picon O. Soft fault detection in cables using the cluster time-frequency domain reflectometry. IEEE Electromagnetic Compatibility Magazine. 2013;**2**(3):70-75

[15] Sallem S, Ravot N et al. Self-adaptive correlation method for soft defect

detection in cable by reflectometry. In: Proceedings of IEEE Sensors. IEEE; 2014. pp. 1-4

[16] Sallem S, Ravot N, et al. Soft defects localization by signature magnification with selective windowing. In: Proceedings of IEEE Sensors. IEEE; 2015. pp. 1-4

[17] Ben Hassen W, Kafal M. Online chafing fault diagnosis and characterization in TP cables based on multi-carrier reflectometry and genetic optimization algorithms. Journal of Cable and Connectivity. 2019;**3**(2):15-21

[18] Schuet S, Dogan AT, Kevin WR. Physics-based precursor wiring diagnostics for shielded-twisted-pair cable. IEEE Transactions on Instrumentation and Measurement. 2015;**64**(12):3420-3432. DOI: 10.1109/TIM.2015.2459052

[19] Schuet S, Timucin D. A model-based probabilistic inversion framework for characterizing wire fault detection using TDR. IEEE Transactions on Instrumentation and Measurement. 2011;**60**(1):88-98

[20] Antoni J, Randall RB. The spectral kurtosis: Application to the vibratory surveillance and diagnostics of rotating machines. Mechanical Systems and Signal Processing. 2006;**20**(2):308-331

[21] Hyun SY, Hong JS, Yun SY, Kim CH, Lee Y. Arc modeling and kurtosis detection of fault with arc in power distribution networks. Applied Sciences. 2022;**12**(6):2777. DOI: 10.3390/app12062777

[22] Taki N, Delpha C, Diallo D, Ben Hassen W, Ravot N. Soft fault diagnosis in wiring networks using reflectometry and principal component analysis. Measurement. 2022;**198**:111378.

DOI: 10.1016/j.measurement.2022.111378

[23] Prieto MD, Cirrincione G, Espinosa AG, Ortega JA, Henao H. Bearing fault detection by a novel condition-monitoring scheme based on statistical-time features and neural networks. IEEE Transactions on Industrial Electronics. 2013;**60**(8):3398-3407

[24] Osman O, Sallem S, Sommervogel L, et al. Distributed reflectometry for soft fault identification in wired networks using neural network and genetic algorithm. IEEE Sensors Journal. 2020;**20**(9):4850-4858

[25] Laib A, Terriche Y, Melit M, Su CL, Mutarraf MU, Bouchekara HREH, et al. Enhanced artificial intelligence technique for soft fault localization and identification in complex aircraft microgrids. Engineering Applications of Artificial Intelligence. 2024;**127**(Part B):107289. DOI: 10.1016/j.engappai.2023.107289

[26] Ben Hassen W, Auzanneau F, Incarbone L, Pérès F, Tchangani AP. Distributed sensor fusion for wire fault location using sensor clustering strategy. International Journal of Distributed Sensor Networks. 2015;**11**(4):538643. DOI: 10.1155/2015/538643

[27] Ben Hassen W, Ravot N, Dupret A, et al. OMTDR-based embedded cable diagnosis for multiple fire zones detection and location in aircraft engines. In: 2017 IEEE Sensors. New York: IEEE; 2017. pp. 1-3

Chapter 5

Aircraft Health Monitoring Systems (AHMS) Applications in Aviation

Melih Cemal Kushan and Seyid Fehmi Diltemiz

Abstract

Aviation holds a significant position in the development and application of health monitoring technologies. In recent years, the integration of Advanced Health Monitoring Systems (AHMS) in aircraft has become increasingly important for ensuring flight safety, operational efficiency, and cost-effective maintenance. This chapter provides an overview of the historical development of AHMS in the aviation sector and discusses the fundamental approaches adopted in aircraft maintenance practices. Particular emphasis is placed on the measurement techniques and sensor systems utilized to monitor the health status of critical components. Furthermore, the chapter explores various data processing and modeling techniques used to analyze the information collected from onboard sensors, including predictive algorithms and condition-based maintenance strategies. By reviewing these technologies and methods, the chapter aims to highlight the growing role of health monitoring in enhancing aircraft reliability, reducing maintenance costs, and supporting proactive decision-making within the aviation industry.

Keywords: structural health monitoring, predictive maintenance, prognostic, aircraft maintenance, digital twin

1. Introduction

Aircraft Health Monitoring Systems (AHMS) have become a critical technology to improve the safety of aircraft, optimize maintenance costs and maximize operational efficiency. AHMS enable real-time monitoring, damage detection and life prediction of aircraft components (engines, wings, fuselage, avionics). It is also very useful tool for the development of preventive maintenance strategies [1]. Another important objective of AHMS is to detect these small changes before failure occurs at the earliest possible opportunity, thereby minimizing downtime, operating costs and maintenance costs, and taking corrective action to reduce the risk of catastrophic failure, injury and even loss of life [2].

In recent years, AHMS technologies have undergone a major transformation with advanced sensor networks, machine learning algorithms and digital twin

IntechOpen

applications. Fiber optic sensors, piezoelectric materials and wireless data transmission have significantly increased the sensitivity and usefulness of AHMS [3]. In addition to improving flight safety, the adoption of SHM in the aviation industry has resulted in significant savings in operating costs by reducing planned maintenance processes, reducing flight postponement rates due to unplanned failures, and preventing further damage by taking corrective action before parts fail [4].

Typical applications of AHMS in aviation include engine performance monitoring, structural crack detection, composite material damage analysis and health assessment of avionics systems [5]. In addition, AHMS, integrated with digital twin technologies, have added a new dimension to failure prediction and maintenance planning by simulating virtual models of aircraft [6].

2. Health monitoring: Definition, history and uses

Briefly, Aircraft Health Monitoring Systems (AHMS) are advanced technology solutions that monitor the condition of structural and mechanical components of aircraft in real time or periodically. By analyzing data collected through sensors, these systems aim to detect potential failures in advance, improve flight safety and optimize maintenance costs.

2.1 The history and important concepts of aircraft health monitoring

Pioneering fields, such as aviation and aerospace, have played a major role in the development of structural health monitoring techniques to the current level, with the main motivating factors such as ensuring flight safety and reducing maintenance costs.

Health monitoring applications in the aerospace industry, as in other areas of technology, have evolved and matured over time with new needs and developing technology. Understanding time-dependent damage progression can be considered as the first important step in structural health monitoring applications. Engineers and academics began to recognize the impact of cyclic stress on the lifespan of components. This understanding led to the development of the S-N curve, which relates the magnitude of cyclic stress to the logarithm of the number of cycles to failure [7]. Initially, the focus was on the safe life approach, which focuses on designing structures to last for a predetermined period without failure. This philosophy manifested itself in structures subjected to repetitive loads, such as railroad bridges and steam boilers, during the steam engine era in the mid-1800s [8, 9].

Another very important building block on the road to AHMS applications is the damage tolerant design philosophy. The evolution of the damage tolerant design philosophy dates back to the mid-twentieth century and marks a significant shift in aerospace engineering practices.

By the 1970s, the limitations of the safe life approach led to the emergence of the damage tolerance philosophy. This new paradigm was introduced to account for the potential growth of existing cracks and defects in materials, rather than focusing solely on preventing their occurrence. Regulatory requirements for damage tolerance analysis were formally integrated into aviation documentation in 1976 and have significantly shaped design standards for aircraft and other heavily loaded aerospace structures [9, 10]. Historical events such as the Antonov An-10 aircraft crash have highlighted the need for robust design methodologies that can accommodate undetected damage and have led to the creation of damage tolerant structures [10]. This philosophy not only considers the initial integrity of a structure but also empha-sizes the importance of monitoring and managing the growth of any damage that may occur during its operational life.

The concept of health monitoring in the aviation industry has evolved sig-nificantly, driven by the continuous pursuit of improved safety and operational efficiency. In the 1970s, pioneering airlines such as British Airways, Air France, and TAP Air Portugal launched the first Flight Data Monitoring (FDM) programs. These programs aim to analyze flight data to enhance operational safety and identify performance-related issues [11].

This early groundwork laid the foundation for contemporary Prognostic Health Management (PHM) systems, reflecting a long-standing commitment to proactive safety measures. As industry progressed, the development of the Aircraft Communications Addressing and Reporting System (ACARS) in the 1980s marked a transformative moment in aircraft monitoring. This automated data link communication system enables the seamless transmission of critical data from the aircraft to airline operations centers, eliminating the need for manual intervention by mechanics [12].

In the early 1990s, these monitoring systems began to gain wider acceptance, reflecting industry-wide recognition of the need for proactive safety measures. Today, the integration of advanced sensor technologies and artificial intelligence into health monitoring systems continues to shape the future of aviation, providing major benefits in the areas of early fault detection and predictive maintenance [13, 14].

Table 1 shows several important events and studies that have played a milestone role in the development of structured health monitoring technology to date.

Health monitoring systems used in aircraft and other aerial platforms can be classified as On-board Health Monitoring Systems (OHMS) and ground-based health monitoring systems (GHMS). Aircraft Health Monitoring Systems (AHMS) utilize a network of sensors installed on critical aircraft components and continuously moni-tor parameters such as engine performance, structural integrity, and various system functions. The collected data is transmitted to GHMS, where ground maintenance teams analyze the information to make informed maintenance decisions. This system enables timely interventions that prevent costly repairs and reduce in-flight failures.

On-board Health Management Systems (OHMS) are critical components in mod-ern aviation systems designed to ensure aircraft safety, reliability, and operational efficiency. These systems employ advanced sensors, data analytics, and predictive technologies to monitor the health of various aircraft systems in real time, enabling proactive maintenance and reducing lifecycle costs [27, 28].

OHMS leverages a network of integrated sensors that continuously collect and analyze operational data from the aircraft during flight. This capability allows maintenance teams to detect potential issues before they evolve into failures, thereby minimizing delays and enhancing passenger safety and comfort [29].

One significant advantage of OHMS is its ability to perform real-time monitoring of aircraft health. Sensor data is instantly transmitted to ground stations or cloud-based platforms for immediate analysis, enabling timely interventions [30]. In other words, OHMS simultaneously generates the database required by GHMS.

The benefits of OHMS include providing a holistic approach to aircraft main-tenance by integrating data from health monitoring systems with maintenance databases and operational programs. Automated systems within OHMS can issue maintenance alerts and safety warnings based on predictive analyses that help inform

Year	Milestone	Institution/ Key contributors	Significance
1860	Metal fatigue theory formulation	August Wöhler	Basis for damage tolerance principles [14].
1921	Vibration analysis fundamentals	Stephen Timoshenko	Enabled dynamic structural analysis [15].
1946	Ultrasonic flaw detection	Firestone	First practical ultrasonic NDT method [16].
1950	Acoustic emission discovery	Josef Kaiser	Real-time crack detection [17].
1955	Experimental modal analysis	Bishop & Gladwell	Vibration-based damage ID [18].
1958	USAF Aircraft Structural Integrity Program (ASIP) was established	United States Air Force	Safe life and damage tolerant design concepts were born [19].
1970	The concept of digital twin was first used	NASA	The concept of digital twin was first used during Apollo 13 mission [20].
1978	Fiber Bragg grating sensors	Butter & Hocker	Distributed strain measurement [21].
1970– 1980	Thermographic inspection using Focal Plane Array	IBM	First practical two-dimensional arrays of infrared detectors to be produced and effectively cooled [22].
1991	The Digital Twin (DT) concept emerged	David Gelernter	The starting era of digital twin [23].
1995	Embedded piezoelectric sensors	Chang F.K., Stanford University	Embedding arrays of piezoelectric sensors within composite laminates during manufacturing [24].
2012	Vehicle Health Management Project	NASA	One of the earliest implementations, when NASA developed prototype digital twins for aircraft engines and structural components to predict component failures before they occurred [25].
2013	ARP6461 Guidelines for Implementation of Structural Health Monitoring on Fixed Wing Aircraft	SAE International	Important industry-wide recommended practice documents specifically for aviation SHM implementation [26].

Table 1.
Key milestones in the development of aircraft health monitoring technology.

both in-flight decisions and ground operations [31]. This consequently reduces aircraft downtime and associated costs [32, 33].

Ground-based health management systems (GHMS) are essential components in modern aviation, providing critical analysis and monitoring capabilities for aircraft operating environments and conducting in-depth analyses of aircraft component health. These systems operate outside the aircraft and utilize advanced data analytics to enhance safety and efficiency throughout the aircraft's lifecycle. Not being constrained by aircraft weight and volume limitations, GHMS systems benefit from powerful data analysis hardware and servers. More complex operations such as health information of the entire fleet, maintenance information, and calculation of efficient flight routes for passenger aircraft are performed within GHMS. It enables maintenance teams to monitor fleet-wide performance and predict potential failures before they occur, reducing unplanned maintenance and operational costs [34, 35].

GHMS collects data from multiple aircraft within a fleet and processes them at maintenance, repair, and overhaul (MRO) centers. This data includes vital indicators such as vibration, temperature, and structural integrity, which are necessary for assessing aircraft health [31].

By applying advanced data analytics techniques, OHMS can identify patterns and anomalies indicating wear or impending failures, allowing for timely maintenance interventions [34, 35]. The transition from reactive maintenance to predictive maintenance is a fundamental aspect of GHMS. These systems analyze historical and current data using complex data analytics to predict potential failures [30, 35]. This proactive approach significantly increases aircraft availability and operational efficiency while minimizing maintenance costs. The success of predictive maintenance largely depends on the quality of the analyzed data, making data integrity a top priority [35, 36].

2.2 Basic approaches used in aircraft health monitoring Modeling

In aircraft health monitoring applications, various models are utilized to make accurate maintenance decisions, avoid unnecessary maintenance costs, and ensure flight safety. The first models were mostly physics-based and began to be used in the 1970s and 1980s. These models were developed based on the principles of materials science, structural mechanics, and fatigue theories [37].

Data-based models became widespread in the late 1990s and especially in the early 2000s. In this period, with the development of sensor technologies, increased data storage capacities and the advancement of machine learning algorithms, it has become possible to extract meaningful maintenance information from large data sets [38]. After the 2010s, hybrid models combining both approaches have gained importance [38, 39].

2.2.1 Physics-based modeling prognostic approach

Physics-based models work with mathematical equations based on the material properties, geometry and dynamic behavior of the structure. By understanding the underlying physics of what happens, engineers can predict potential failures and maintenance needs before they occur, improving safety and efficiency in operations [39–41]. Finite element analysis (FEA) and modal analysis are the most commonly used tools in this approach [1].

Its advantages include the ability to operate with limited data, explain fusional processes, and model long-term material behavior. However, it also has disadvantages, such as requiring high computational power and facing difficulties in modeling complex systems, where parameter uncertainties—such as material properties and temperature conditions—can lead to significant errors.

2.2.2 Data-based modeling prognostic approach

Data-based modeling focuses on leveraging historical data collected from aircraft systems to identify patterns and anomalies [39–41]. The data-driven model detects anomalies or damage by analyzing sensor data (accelerometer, strain gauge, acoustic emission, etc.) obtained from the structure with machine learning (ML) and statistical methods. These models are based on the detection of irregularities in the data instead of physical equations. This method provides real-time insights and predictive analytics using algorithms that learn from existing data. It enables proactive maintenance before potential damage occurs. This results in lower failure rates, shorter maintenance times and safer operation. Data-based models are particularly effective in scenarios where physical models may be too complex or costly to develop [41]. Its advantages are that it does not require complex physical models, provides highly accurate results in case of large data sets, and is suitable for real-time monitoring. The disadvantages are that it requires high quality and abundant training data, the difficulty of physical explanation of the results due to its "black box" structure, and the limited extension of the model to different structures.

2.2.3 Hybrid approach

Hybrid SHM models combine the advantages of data-driven and physics-based approaches to provide more reliable and interpretable results. For example, hybrid models can use data to refine and calibrate physical models, improving their predictive capabilities while at the same time ensuring they are grounded in the reality of observed performance [39, 40]. These models are becoming increasingly common, especially in high-risk areas such as aviation, energy and defense industries. The advantages are high accuracy, less data requirement and interpretability. Disadvantages include the need for expertise, high computational power and the difficulty of compatibility of Physics and Data models.

2.2.4 Digital twin approach

A digital twin is a virtual copy of a physical asset or system (airplane, bridge, wind turbine, etc.) fed with real-time data. Digital twins used in AHMS simulate the current state of the structure, enabling failure prediction, optimization and development of maintenance strategies.

In the aviation industry, digital twins are used to improve maintenance processes and increase the safety and efficiency of operations. By enabling the simulation of various scenarios, they help predict wear and tear on components and facilitate proactive maintenance strategies.

By integrating digital twin technology into avionics processes, manufacturers such as Boeing have underscored its importance in modern aviation management. As technology continues to evolve, the potential for digital twins to create comprehensive models spanning entire fleets or organizations will further revolutionize the aviation industry, providing real-time visibility and operational insights on a much larger scale [42–45]. The basic components of a digital twin are the physical system (real structure equipped with sensors), the virtual model, data communication and algorithms that generate predictions.

The advantages are real-time monitoring capability, failure prediction, cost savings, and no "black box" problem thanks to physical models. The disadvantages are high computational power, data security risks and difficulty of integration.

2.3 Measurement techniques used in aircraft health monitoring systems

For a typical AHM system to function properly, a sensor network that measures the required health data, a data acquisition and analysis unit that collects the data from the sensors and analyzes the necessary data in real time, user interfaces designed for the pilot and maintenance team, and algorithms that analyze the data and report the current status and future maintenance recommendations.

In order for the AHM system to fulfill these tasks, it is first necessary to know the stresses and environmental conditions to which the aircraft is exposed during its use. For this purpose, a large number of different purpose-built sensors are used [20]. The main sensor types used in AHMS are given in **Table 2**.

2.4 Application areas of health monitoring systems in aircraft

Various factors, such as the advancement of aircraft designs, innovations in sensor technology, and the development of aircraft maintenance concepts over the years, have enabled the use of health monitoring applications in different parts of aircraft over the years. In this chapter, the main areas where AHMS is used are aircraft structures, gas turbines and accessory units, gearboxes, hydraulic equipment and landing gear.

Sensor type	Measurement characteristics	Typical application examples
Strain gauges	Measures and monitors strain on the object resulting from applied force or temperature	Airframe structural parts such as longerons, bulkheads
Load Cells	Used for measuring tension, compression, pressure or torque	Actuators
Accelerometer	Measures dynamic or static acceleration force in various directions	Airframe, landing gears
Temperature Sensors	Measures temperatures, usually by thermocouples, Infrared sensors or RTDs	Air inlet, compressors and turbines
Fiber Optic Sensors	Measures stress, strains or temperature	Airframe structural parts, cylinders
Acoustic Emission Sensor	Measures the current state of materials under stress. Used to dynamically detect the growth of cracks or corrosion	Airframe structural parts
Vibration Meters	Measures high or low frequency vibrations of the structure	Gear boxes, turbine casings
Linear Variable Differential Transformer (LVDT)	Measures the linear position of the object	Flap actuators
Tiltmeter	Measures the inclined angle of the object	Determination of aircraft, airframe structural parts
G - level Indicator	Measure gravitational forces	Landing gears
Pressure Transducers		Compressors, fans, air inlet pressures, fluid lines
Proximity Sensors	Measures the distance of the objects	Determining turbine shafts, gear shafts revolution speed
Piezoceramic sensors	Measures strain and vibrations	Embedded in structural composite wing panels

Table 2.
Different types of sensors used in AHMS for data collection and their example applications.

2.4.1 Aircraft structural parts health monitoring

The first examples of health monitoring applications in the aerospace industry were carried out on aircraft structural parts. The beginning of these applications can be traced back to the damage tolerance approach design philosophy of the US Air Force, which is also mentioned in history. In the damage tolerance design philosophy, a certain amount of damage is allowed to occur and progress throughout the lifetime of aircraft components to reduce their weight and maximize their performance. The speed at which these damages progress and their critical dimensions are determined at the design stage through calculations and tests. This makes it possible to work with parts that have the exact structural strength required by the design, rather than using heavy and over-engineered parts that are so heavy that they will never fail during their lifetime [19].

Aircraft components, such as spars, longerons and bulkheads, are critical components that provide the main structural integrity of the aircraft. These structural elements are vital in aircraft design as they enable the airframe to withstand various aerodynamic and operational stresses while maintaining the desired shape and function.

The main damages in aircraft structural components that are monitored using structural health monitoring techniques are fatigue cracks, corrosion and wear. Fatigue cracks are seen especially in areas of stress concentration, such as rivet holes, bolt thread bottoms and sharp corners on structural parts. Corrosion damage is generally observed in parts that are exposed to the atmospheric sun, moisture, salt water industrial atmosphere. Protective coatings and paints are used to prevent corrosion in these parts. The primary function of these protective layers is to act as a barrier, preventing the corrosive environment from interacting with the underlying metallic material.

When one thinks of Structural Health Monitoring (SHM), the Aircraft Structural Integrity Program (ASIP) comes to mind. Over the decades, the program has evolved significantly, adapting to the increasing complexity of modern aircraft designs and incorporating advanced methodologies such as damage tolerance and fracture control technologies [46, 47].

Today, ASIP uses advanced health monitoring applications that utilize real-time data to assess aircraft structural integrity, enabling proactive maintenance strategies that significantly improve aviation safety standards [46].

2.4.2 Gas turbine engine health monitoring

Gas turbine engines consist of several key components that work together to produce thrust and power. The primary parts of a gas turbine include the compressor, combustion chamber, turbine and in some cases a power turbine. These elements are essential for engine operation and efficiency [48].

Gas turbine engine monitoring techniques are critical to ensure the reliability and efficiency of engine operations. These methods primarily focus on detecting early signs of failures and assessing the overall health of the engine.

The development of Engine Health Monitoring Systems (EHMS) in aviation can be traced back to their first use on military helicopters in the late 1990s and early 2000s. These early systems laid the groundwork for the more sophisticated technologies that would follow. Over time, advances in computing power and forecasting algorithms have significantly improved the accuracy and efficiency of these monitoring systems, enabling a more precise collection of data based on past performance and operational use cases [49].

Aircraft Engine Health Monitoring Systems (EHMS) provide numerous benefits that significantly improve the efficiency and reliability of aviation operations.

One of the most important benefits of Aircraft Engine Health Monitoring Systems (EHMS) is their contribution to flight safety. These systems continuously monitor complex engine parameters, enabling early detection of potential problems before they become serious flight safety issues. This proactive approach increases confidence in safety practices among aviation operators and helps prevent unexpected downtime caused by engine-related issues, ultimately improving operational reliability [50, 51].

EHMS can reduce maintenance costs by up to 30% by helping to detect and prevent maintenance issues before they lead to failure [12, 52].

Integration of advanced technologies, including machine learning and historical data analysis, enables maintenance teams to accurately predict engine maintenance requirements and efficiently plan interventions, minimizing unscheduled maintenance events and increasing aircraft availability [53, 54].

These systems help aviation companies effectively manage operational risks. EHMS helps airlines comply with regulatory standards and extends the life of their aircraft fleet [55, 56].

AHMS adoption leads to an increase in overall operational efficiency. By optimizing maintenance strategies through advanced algorithms, airlines can reduce downtime and improve scheduling, which ultimately increases the efficiency of fleet operations [57, 58].

As a result, the aviation industry benefits from a comprehensive approach that not only enhances safety and reliability but also improves the economic viability of airline operations [54, 59].

Modern gas turbine health monitoring systems utilize data from a variety of engine-mounted sensors to provide early warnings of potentially dangerous conditions [60]. These systems can operate either in "on-line" mode, which analyzes in real time during the flight, or in "off-line" mode, which allows a comprehensive review of the recorded data after the operation [61].

The use of sensor data requires relatively few detailed theoretical system models, simplifying condition assessment and making it easier for engineers to effectively monitor and manage engine performance [60].

2.4.2.1 Temperature monitoring

Temperature monitoring is a vital aspect of gas turbine engine health assessment. For example, an excessive rise in turbine inlet temperature (TIT) or exhaust outlet temperature (EGT) could indicate permanent damage to the engine and a major failure that puts flight safety at risk. Methods for monitoring temperature typically involve the collection of multiple sensor readings that reflect the operating state of the motor [56, 59].

Given the high ambient temperature, especially for hot section temperature readings, special sensors must be used to accurately measure temperatures without being adversely affected by extreme conditions [62].

2.4.2.2 Vibration-based monitoring

Vibration measurements serve as a fundamental tool for global monitoring of the health status of a gas turbine. The principle underlying this technique is that the dynamics of a gas turbine system change when faults begin to occur, resulting in altered vibration patterns compared to those observed in the healthy state [63, 64]. For example, a crack in the compressor blade or an imbalance caused by even a small part breakage is quickly detected by vibration sensors.

Recent developments show that gas turbine manufacturers are increasingly adopting data-driven vibration-based condition monitoring approaches that improve reliability and availability without relying on complex physics-based models that require numerous assumptions [63].

2.4.2.3 Oil analysis

Oil system monitoring has been one of the oldest and most established techniques in gas turbine engine monitoring. While it has a long history, advances in technology have led to improved methodologies that provide deeper insights into engine performance and condition [65]. For example, in the event of wear in the bearings on the engine, small metal particles mixed into the oil are detected, and further damage to the bearing or engine is prevented [61].

2.4.2.4 Gas turbine performance monitoring

Gas turbine performance monitoring serves as an important tool for early detection of failures. By developing models based on engine data, both linear and non-linear relationships specific to particular engines can be captured. This allows manufacturers to detect defects early during pass tests, significantly reducing the risk of complete component failure. For example, compressor compression ratio is a good performance metric, and a decrease in this value indicates damage to the compressor, such as wear, etc. [63].

Smart embedded sensing technologies further support these monitoring efforts, facilitating smooth operation and improved reliability [66].

2.4.2.5 Life consumption monitoring

Gas Turbine Engine Life Consumption Monitoring is an effective way to prevent in-use component failures. Historically, maintenance strategies were based on fixed time, but modern practices favor condition-based approaches that utilize life monitoring systems specific to each aircraft engine. For example, engine life under idle or after burner (AB) conditions is recorded on an aircraft-by-aircraft basis. By calculating the effect of each regime on engine life differently, the wear level of the engine of each aircraft in the fleet is calculated, and maintenance periods are determined accordingly. These systems have been particularly successful in extending the service life of aging fleets by automating the measurement of engine life [61, 67].

2.4.3 Accessories and auxiliary systems health monitoring

Health monitoring techniques for aircraft gearboxes, landing gear and hydraulic equipment are essential to ensure safety, reliability and operational efficiency in aviation. These components are essential for the proper functioning of various aircraft systems, with gearboxes enabling power transmission, landing gear ensuring safe take-off and landing, and hydraulic systems controlling critical functions. These systems are integral to various functions, including landing gear operation and control surfaces, and rely on the efficient transmission of energy through high-pressure fluids. However, the complexity of hydraulic systems requires meticulous maintenance and monitoring to reduce the risks associated with potential failures.

The importance of effective health monitoring is manifested in its ability to detect potential failures early, thereby minimizing maintenance costs, reducing unplanned downtime and improving overall aircraft performance. Various techniques are used to assess the condition of accessories and auxiliary systems, leveraging advanced technologies for predictive maintenance.

2.4.3.1 Vibration analysis

Vibration analysis is a widely adopted technique for monitoring the operating condition of gearboxes. It involves measuring the vibration patterns produced by the gearbox during operation. When components are in good condition, they produce specific vibration patterns. However, as wear and tear occur on bearings and gears, these patterns change, which is a precursor to potential failures such as misalignment, imbalance or bearing failures [68, 69].

Vibration analysis is based on collecting vibration readings using sensors and then analyzing this data on established bases to identify anomalies. Serious failures in hydraulic units, such as gear breakage, bearing failure, etc., often manifest themselves in the form of sudden increases in vibration values.

2.4.3.2 Thermography

Thermography or thermal imaging uses infrared cameras or thermocouples through the fluid line to detect temperature variations inside gearboxes. Abnormal heat values often indicate problems such as excessive friction or mechanical wear [70, 71]. By monitoring temperature changes, technicians can diagnose potential problems before they lead to catastrophic failures.

2.4.3.3 Oil analysis

Oil analysis is another vital monitoring technique. Regular sampling and analysis of the oil used in gearboxes can reveal critical information about the condition of internal components. The analysis focuses on detecting contaminants, wear particles and oil quality degradation that could indicate problems such as bearing wear or inadequate lubrication [72].

This proactive approach ensures that maintenance is performed at optimal intervals, prolonging the life and safe operation of hydraulic components such as gearboxes, bearings and pumps.

2.4.3.4 Load and stress monitoring

Various methods have been developed to monitor the health of landing gears, often involving various sensors and collection systems. These methods enable the structural integrity of the landing gear to be analyzed in real time, allowing early detection of potential problems such as wear and tear [72, 73].

Common approaches include direct load measurements, strut detection and weight/balance calculations, which collectively contribute to a comprehensive understanding of the condition of the landing gear [74, 75].

Structural Health Monitoring (SHM) systems play a crucial role in assessing the condition of landing gear. These systems use advanced sensors integrated into the landing gear assembly to monitor loads and detect damage escalation in real time. Implementing SHM not only improves safety by facilitating condition-based maintenance (CBM) but also extends the service life of landing gear components [75].

By continuously monitoring the health status of the landing gear, potential failures can be detected at an early stage, and thus maintenance actions can be planned before failures occur [73, 76].

3. Conclusions

Maximizing flight safety has been an ongoing and never-ending demand in the aviation industry since its inception. To meet this demand, Aircraft Health Monitoring Systems (AHMSs) have been continuously evolving for many years. This historical development process not only emphasizes the importance of learning from past experiences but also demonstrates how technology has constantly evolved to meet the growing demands of air transportation and safety regulations.

Innovations in sensor technology and computational models have paved the way for powerful computing and health monitoring systems, such as digital twins. As we enter the era of big data, the integration of health monitoring systems into maintenance strategies is progressing at an accelerating pace.

Author details

Melih Cemal Kushan* and Seyid Fehmi Diltemiz
Eskisehir Osmangazi University, Eskisehir, Turkiye

*Address all correspondence to: erzesk@gmail.com

IntechOpen

© 2025 The Author(s). Licensee IntechOpen. This chapter is distributed under the terms of the Creative Commons Attribution License (http://creativecommons.org/licenses/by/4.0), which permits unrestricted use, distribution, and reproduction in any medium, provided the original work is properly cited. (cc) BY

References

[1] Farrar CR, Worden K. Structural Health Monitoring. United Kingdom: Wiley; 2012. DOI: 10.1002/9781118443118

[2] Ross RW. Integrated vehicle health management in aerospace structures. In: Structural Health Monitoring (SHM) in Aerospace Structures. United Kingdom: Elsevier; 2016. pp. 3-31. DOI: 10.1016/B978-0-08-100148-6.00001-9

[3] Giurgiutiu V. Introduction. In: Structural Health Monitoring with Piezoelectric Wafer Active Sensors. Amsterdam: Elsevier; 2014. pp. 1-19. DOI: 10.1016/B978-0-12-418691-0.00001-0

[4] Boeing at Forefront of Revolution in Predictive Maintenance. United States: Boeing Services; n.d. Available from: https://services.boeing.com/news/boeing-at-forefront-of-revolution-in-predictive-maintenance [Accessed: April 11, 2025]

[5] Ihn J-B, Chang F-K. Pitch-catch active sensing methods in structural health monitoring for aircraft structures. Structural Health Monitoring. 2008;7:5-19. DOI: 10.1177/1475921707081979

[6] Tao F, Zhang H, Liu A, Nee AYC. Digital twin in industry: State-of-the-art. IEEE Transactions on Industrial Informatics. 2019;15:2405-2415. DOI: 10.1109/TII.2018.2873186

[7] Lisitsin O. Damage-Tolerant Design: A Few Words about Safely Damaged Structures. Ukraine: Engre; n.d. Available from: https://engre.co/blogs/articles/damage-tolerant-design-a-few-words-about-safely-damaged-structures/ [Accessed: April 20, 2025]

[8] DTD Handbook. Introduction. United States: University of Dayton Research Institute; n.d. Available from: https://www.afgrow.net/applications/dtdhandbook/sections/page1_0.aspx [Accessed: April 20, 2025]

[9] Reddick HK Jr. Safe-life and damage-tolerant design approaches for helicopter structures. In: NASA Langley Research Center Failure Anal and Mech of Failure of Fibrous Composite Struct. United States: NASA; 1983

[10] Chisholm SA, Rufin AC, Chapman BD, Benson QJ. Forty years of structural durability and damage tolerance at Boeing commercial airplanes. Boeing Technical Journal. 2016:1-24

[11] The Impact of Global Regulations on Aviation Compliance. eLeaP®; n.d. Available from: https://quality.eleapsoftware.com/the-impact-of-global-regulations-on-aviation-compliance/ [Accessed: April 20, 2025]

[12] Raval C. Importance of Health Monitoring System for Aircraft. United States: Arrow Company; n.d. Available from: https://www.einfochips.com/blog/importance-of-health-monitoring-system-for-aircraft/ [Accessed: April 20, 2025]

[13] Robins M. September/October 2023 - Aircraft Health Monitoring. Avionics Digital Edition. Unites States: Avionics International; n.d. Available from: https://interactive.aviationtoday.com/avionicsmagazine/september-october-2023/aircraft-health-monitoring/ [Accessed: April 20, 2025]

[14] Wöhler A. Versuche über die Festigkeit der Eisenbahnwagenachsen.

Zeitschrift Für Bauwesen.
1860;**10**:160-161

[15] Timoshenko SP LXVI.
On the correction for shear of the
differential equation for transverse
vibrations of prismatic bars. The
London, Edinburgh, and Dublin
Philosophical Magazine and Journal
of Science. 1921;**41**:744-746. DOI:
10.1080/14786442108636264

[16] Firestone FA. The supersonic
reflectoscope for interior inspection.
Metal Progress. 1945;**48**:505-512

[17] Kaiser J. Untersuchungen über
das Auftreten von Geräuschen beim
Zugversuch. Munich: Technische
Universität München; 1950

[18] Donohue BRE, Johnson DC.
The Mechanics of Vibration. United
Kingdom: Cambridge University Press;
1962

[19] Ball DL, Burt RJ. Evolution of the
USAF aircraft structural integrity
program. In: USAF Aircraft Structural
Integrity Program Conference. San
Antonio Texas: Lockheed Martin
Corporation; 2016

[20] Shah V, Suthar A. Possibilities
with digital twin. Digital Twin
Technology: Fundamentals and
Applications. 2022:219-232. DOI:
10.1002/9781119842316.CH14

[21] Hocker GB, Butter CD. Fiber
optics strain gauge. Applied Optics.
1978;**17**(18):2867-2869. DOI: 10.1364/
AO.17.002867

[22] Khodayar F, Sojasi S,
Maldague X. Infrared thermography
and NDT: 2050 horizon. Quantitative
InfraRed Thermography
Journal. 2016;**13**:210-231. DOI:
10.1080/17686733.2016.1200265

[23] Bisanti GM, Mainetti L,
Montanaro T, Patrono L, Sergi I. Digital
twins for aircraft maintenance and
operation: A systematic literature review
and an IoT-enabled modular architecture.
Internet of Things. 2023;**24**:100991. DOI:
10.1016/J.IOT.2023.100991

[24] Chang F-K. Built-in damage
diagnostics for composite structures. UK:
Woodhead Publishing Limited; 1995

[25] Glaessgen EH. The digital twin
paradigm for future NASA and U.S.
Air Force Vehicles. In: NASA Technical
Reports Server (NTRS). United States:
NASA; n.d. Available from: https://
ntrs.nasa.gov/citations/20120008178
[Accessed: April 21, 2025]

[26] Aerospace Industry Steering
Committee on Structural Health.
ARP6461: Guidelines for Implementation
of Structural Health Monitoring on
Fixed Wing Aircraft. United States:
SAE International; n.d. Available from:
https://www.sae.org/standards/content/
arp6461/ [Accessed: April 21, 2025]

[27] Olaganathan R, Miller M,
Mrusek B. Managing safety risks in
airline maintenance outsourcing.
International Journal of Aviation,
Aeronautics, and Aerospace. 2020;**7**(1):1-
21. DOI: 10.15394/ijaaa.2020.1435

[28] GE Aerospace. Aircraft Health
Management Unit. United States;
2016. Available from: https://www.
geaerospace.com/sites/default/files/2022-
01/AircraftHealthManagementUnit-
DataSheet.pdf [Accessed: October 30,
2025]

[29] Surender S. How Aircraft Health
Monitoring Systems are Shaping
the Future of Aerospace Industries.
Markets and Markets Corp.; n.d.
Available from: https://www.
marketsandmarkets.com/blog/AD/

How-Aircraft-Health-Monitoring-Systems-Revolutionizing-The-Future-Of-Aerospace-Industries [Accessed: April 20, 2025]

[30] Logue J. Digital Transformation in Aircraft Maintenance: The Role of Data and Analytics. United States: Logue Aviation: Expert Aircraft Consulting & Maintenance Services; n.d. Available from: https://LogueaviationCom/

[31] Onboard Maintenance Systems. United States: Honeywell; n.d. Available from: https://aerospace.honeywell.com/us/en/products-and-services/products/cabin-and-cockpit/data-gateways/onboard-maintenance-systems [Accessed: April 20, 2025]

[32] Bellamy W. December 2016/January 2017 - An Insider Look at How FedEx is Digitally Managing Airplane Health. United States: Avionics Digital Edition; n.d. Available from: https://interactive.aviationtoday.com/avionicsmagazine/december-2016-january-2017/an-insider-look-at-how-fedex-is-digitally-managing-airplane-health/ [Accessed April 20, 2025]

[33] McHugh B. The Untapped Potential of PHM Solutions. Collins Aerospace Corp.; n.d. Available from: https://connectedaviationtoday.com/the-future-of-aircraft-maintenance-untapped-potential-in-phm-solutions/ [Accessed: April 20, 2025]

[34] Tang L, Saxena A, Younsi K. Prognostics and health management for electrified aircraft propulsion: State of the art and challenges. Journal of Engineering for Gas Turbines and Power. 2025;147. DOI: 10.1115/1.4066598

[35] Airplane Health Management (AHM). Boeing Services. United States: Boeing Aerospace Corp.; n.d. Available from: https://services.boeing.com/maintenance-engineering/maintenance-optimization/airplane-health-management-ahm [Accessed: April 20, 2025]

[36] Soaring to New Heights - The Expanding Aircraft Health Monitoring Systems Market. Market Research Intellect Corp. Available from: https://www.marketresearchintellect.com/blog/soaring-to-new-heights-the-expandingaircraft-health-monitoring-systemsmarket/ [Accessed: April 20, 2025]

[37] Jardine AKS, Lin D, Banjevic D. A review on machinery diagnostics and prognostics implementing condition-based maintenance. Mechanical Systems and Signal Processing. 2006;**20**:1483-1510. DOI: 10.1016/J.YMSSP.2005.09.012

[38] Kai G, Daigle MJ, Abhinav S, Indranil R, Shankar S, Celaya JR. Prognostics : The Science of Making Predictions. United States: CreateSpace Independent Publishing Platform; 2017

[39] Fu S, Avdelidis NP. Prognostic and health management of critical aircraft systems and components: An overview. Sensors (Basel). 2023;**23**:8124. DOI: 10.3390/S23198124

[40] Khan S, Yairi T, Tsutsumi S, Nakasuka S. A review of physics-based learning for system health management. Annual Reviews in Control. 2024;**57**:100932. DOI: 10.1016/J.ARCONTROL.2024.100932

[41] An D, Kim NH, Choi J-H. Practical options for selecting data-driven or physics-based prognostics algorithms with reviews. Reliability Engineering and System Safety. 2015;**133**:223-236. DOI: 10.1016/j.ress.2014.09.014

[42] Digital Twins. Transforming Aircraft Lifecycle Management. United States:

Saabrds Corp.; n.d. Available from: https://saabrds.com/digital-twins-revolutionizing-aircraft-lifecycle-management/ [Accessed: April 23, 2025]

[43] The Game-Changing Influence of Digital Twins on Cybersecurity. United States: CounterCraft; n.d. Available from: https://www.countercraftsec.com/blog/the-game-changing-influence-of-digital-twins-on-cybersecurity/ [Accessed: April 23, 2025]

[44] Digital Twin. ACRP Transformative Tech. United States: Transportation Research Board Academy; n.d. Available from: https://crp.trb.org/acrptransformativetech/technology-focus-articles/digital-twin/ [Accessed: April 23, 2025]

[45] Kritzinger W, Karner M, Traar G, Henjes J, Sihn W. Digital twin in manufacturing: A categorical literature review and classification. IFAC-PapersOnLine. 2018;**51**:1016-1022. DOI: 10.1016/J.IFACOL.2018.08.474

[46] Negaard GR. The History of the Aircraft Structural Integrity Program. Scientific Report. United States: US Airforce Aerospace Structures Information and Analysis Center; 1980

[47] Wells HM JR, King TT. Air Force Aircraft Structural Integrity Programo: Airplane Requirements. United States: Ohio; n.d

[48] Gas Turbine: Parts, Working, Types, Advantages, and Applications. India: Testbook; n.d. Available from: https://testbook.com/mechanical-engineering/gas-turbine [Accessed: April 21, 2025]

[49] Rahim MA, Rahman MM, Islam MS, Muzahid AJM, Rahman MA, Ramasamy D. Deep learning-based vehicular engine health monitoring system utilising a hybrid convolutional neural network/bidirectional gated recurrent unit. Expert Systems with Applications. 2024;**257**:125080. DOI: 10.1016/J.ESWA.2024.125080

[50] Waters N. Engine Health Managment. United Kingdom: Ingenia; n.d. Available from: https://www.ingenia.org.uk/articles/engine-health-managment/ [Accessed: April 21, 2025]

[51] Kabashkin I, Perekrestov V, Tyncherov T, Shoshin L, Susanin V. Framework for integration of health monitoring systems in life cycle management for aviation sustainability and cost efficiency. Sustainability. 2024;**16**:6154. DOI: 10.3390/SU16146154

[52] Rappe V. What Type of Aerospace Temperature Sensors Are There? United States: Ametek Sensors; n.d. Available from: https://www.ameteksfms.com/pressreleases/articles/2021/august/aerospace-temperature-sensors [Accessed: April 21, 2025]

[53] Kabashkin I, Shoshin L. Artificial intelligence of things as new paradigm in aviation health monitoring systems. Future Internet. 2024;**16**:276. DOI: 10.3390/FI16080276

[54] Kabashkin I. AI and evolutionary computation for intelligent aviation health monitoring. Electronics. 2025;**14**:1369. DOI: 10.3390/ELECTRONICS14071369

[55] Wang L, Zhao X, Pham H. Novel formulations and metaheuristic algorithms for predictive maintenance of aircraft engines with remaining useful life prediction. Reliability Engineering and System Safety. 2025;**261**:111064. DOI: 10.1016/J.RESS.2025.111064

[56] Engine Health and Vibration Monitoring. United States: Meggitt; n.d. Available from: https://www.meggitt.

com/products-services/engine-health-and-vibration-monitoring/ [Accessed: April 21, 2025]

[57] AC 33.28-1 - Compliance Criteria for 14 CFR §33.28, Aircraft Engines, Electrical and Electronic Engine Control Systems. United States: Federal Aviation and Administration; n.d. Available from: https://www.faa.gov/regulations_policies/advisory_circulars/index.cfm/go/document.information/documentID/22925 [Accessed: April 21, 2025]

[58] Schmitz S. Using Engine Condition Trend Monitoring As a Standard. United States: Duncan Aviation; n.d. Available from: https://www.duncanaviation.aero/intelligence/using-engine-condition-trend-monitoring-as-a-standard [Accessed: April 21, 2025]

[59] Engine Health Monitoring. United States: Standart Aero Company; n.d. Available from: https://standardaero.com/customsolutions/enginehealthmonitoring/ [Accessed: April 21, 2025]

[60] Types of Gas Turbines. United States: NASA Glenn Research Center; n.d. Available from: https://www.grc.nasa.gov/www/k-12/airplane/trbtyp.html [Accessed April 21, 2025]

[61] Yousif S, Alnaimi F, Thiruchelvam S. Gas turbine performance monitoring and operation challenges: A review. Gazi University Journal of Science. 2023;**36**:154-171. DOI: 10.35378/GUJS.948875

[62] Minimising The Cost of Gas Turbine Failure through Performance Monitoring and Early Issue Detection. United Kingdom: Hawkins Forensic Investigation; n.d. Available from: https://www.hawkins.biz/insight/minimising-the-cost-of-gas-turbine-

failure-through-performance-monitoring-and-early-issue-detection/ [Accessed: April 21, 2025]

[63] Tauber T. AIR 1828 - A Guide to Gas Turbine Engine Oil System Monitoring. SAE Technical Papers. United States: Society of Automotive Engineers (SAE); 1983. DOI: 10.4271/831477

[64] Murugan M, Walock M, Ghoshal A, Knapp R, Caesley R. Embedded sensing for gas turbine engine component health monitoring. In: Vertical Flight Society's 76th Annual Forum and Technology Display. United States: Vertical Flight Society Organization; 2020. DOI: 10.4050/F-0076-2020-16339

[65] Gabraeel N, Lieuwen T, et al. Realtime Health Monitoring of Gas Turbine Components Using Online Learning and High Dimensional Data. United States: Georgia Tech. University; 2018

[66] Meher CB, Cullen JP. Integration of condition monitoring Technologies for the Health Monitoring of gas turbines. In: International Gas Turbine and Aeroengine Congress and Exposition. Cologne: ASME; 1992

[67] Gas Turbine Monitoring. United States: Kistler Company; n.d. Available from: https://www.kistler.com/US/en/gas-turbine-monitoring/C00000025 [Accessed: April 21, 2025]

[68] CM of Gear Boxes – Reliability Edge. India: Edge Reliability Corp.; n.d. Available from: https://edgereliability.com/service/cm-of-gear-boxes/ [Accessed: April 21, 2025]

[69] Predictive Maintenance for Crane Motors and Gearboxes. United States: Banner Engineering Corp.; n.d. Available from: https://www.bannerengineering.com/us/en/

solutions/iiot-data-driven-factory/
predictive-maintenance-of-motors-
and-gearboxes-on-large-cranes.
html?srsltid=AfmBOoqj6W_UFQUAG3
G1Y7DLjLr-kj5NvPYRwXSkYF-
gGoF3tCwhPgq8 [Accessed: April 21,
2025]

[70] Wawrin M. Condition Monitoring
Techniques You Must Know. Canada:
Click Maint Corp.; n.d. Available from:
https://www.clickmaint.com/blog/
condition-monitoring-techniques
[Accessed: April 21, 2025]

[71] Suresh S, Naidu V. Gearbox
health condition monitoring: A brief
exposition. United States: Control and
Data Fusion e-Journal. n.d. Available
from: https://www.researchgate.net/
publication/369113235_Gearbox_
Health_Condition_Monitoring_A_brief_
exposition [Accessed: April 21, 2025]

[72] Skorupka Z, Tywoniuk A. Health
monitoring in landing gears. Journal of
KONES. 2019;**26**:167-174. DOI: 10.2478/
kones-2019-0020

[73] Hosseini M. Predictive Maintenance
in Aircraft (Aviation) Industry. United
States: Arshon Inc. Blog; n.d. Available
from: https://arshon.com/blog/
predictive-maintenance-in-aircraft-
aviation-industry/ [Accessed: April 21,
2025]

[74] Forrest CB, Forrest C, Wiser D.
Landing gear structural health
monitoring system. In: (EWSHM 2018)
9th European Workshop on Structural
Health Monitoring (EWSHM 2018);
10-13 July 2018; Manchester, UK:
(EWSHM 2018). United Kingdom; n.d.
Available from: https://www.ndt.net/
search/docs.php3?id=23297 [Accessed:
April 21, 2025]

[75] Forrest C, Forrest C, Wiser D.
Landing gear structural health

monitoring (SHM). Procedia Structural
Integrity. 2017;**5**:1153-1159. DOI:
10.1016/J.PROSTR.2017.07.025

[76] Jobmann J, Thielecke F. Model-based
loads observer approach for landing
gear remaining useful life prediction.
PHM Society European Conference.
2024;**8**:984-994. DOI: 10.36001/
PHME.2024.V8I1.4104

www.ingramcontent.com/pod-product-compliance
Lightning Source LLC
Chambersburg PA
CBHW081336190326
41458CB00018B/6022